Erwin Schanda Physical Fundamentals of Remote Sensing

With 102 Figures and 14 Tables

Springer-Verlag
Berlin Heidelberg New York Tokyo

Professor Dr. ERWIN SCHANDA
University of Bern
Institute of Applied Physics
Sidlerstraße 5
3012 Bern, Switzerland

ISBN 3-540-16236-4 Springer-Verlag Berlin Heidelberg New York Tokyo
ISBN 0-387-16236-4 Springer-Verlag New York Heidelberg Berlin Tokyo

Libray of Congress Cataloging-in-Publication Data. Schanda, Erwin. Physical fundamentals of remote sensing. Includes index. 1. Remote sensing I. Title. G70.4.S34 1986   621.36'78
85-31735

© by Springer-Verlag Berlin Heidelberg 1986
Printed in Germany

Typesetting: K + V Fotosatz GmbH, Beerfelden
Offsetprinting and bookbinding: Beltz Offsetdruck, Hemsbach
2132/3130-543210

Preface

Ten years ago the author, together with eight co-authors, edited a textbook *Remote Sensing for Environmental Sciences* within the series on Ecological Studies of Springer-Verlag. At that time there were not yet many books available on remote sensing. The decade that has elapsed was marked by a spectacular development in this field. This development took place in many directions: by widening the areas of application, by improvements of the methods and the sensors, by the introduction of new versatile platforms, but also by deepening the knowledge of the theoretical foundations. This evolution improved the ability to explain the interaction between electromagnetic radiation and natural objects, which, in its turn, allowed for better modelization and for the creation of refined mathematical tools in the processing of remotely sensed data and in the determination of the physical status of remote objects. The community of research workers engaged in development and use of remote sensing methods changed accordingly from a modest group of scientists in the early 1970's to a considerable branch of specialized and interdisciplinary activity. The training of students had to be adapted to cope with the increasing number of people entering this new field and with the increasing quality of the material to be presented. At present scientists exchange their results at many international conferences each year, and the annual production of publications in various related scientific journals amounts to several thousands of pages in addition to the growing number of monographs and manuals. One could ask for what purpose should one publish another book?

According to the author's experience, however, there is a serious neglect of systematic study and education about important physical fundamentals.

The present text is the result of the author's graduate course as it evolved during the last 6 years. This subject, under the title *Physikalische Grundlagen der Fernerkundung für Naturwissenschaftler* was presented as a course for the first time during the summer term of 1979 at the University of Bern, Switzerland. The material was considerably supplemented and arranged in a shape very much like the present text when the lectures were presented at the Université Paul Sabatier, Toulouse, France, during the academic years 1982/83 and 1983/84 under the title *Interaction du rayonnement electromagnétique et des milieux naturels.*

Within the limits of the present volume, no complete treatment of all significant physical fundamentals of remote sensing is possible. Therefore, topics were selected which are, according to the personal judgement and taste of the author, most important for the comprehension of physical relationships relevant in the methods of remote sensing and their applications to the environment. However, neither mathematical methods, such as, for example for data and image processing, nor the environmental phenomena and processes were intended to be treated within this text.

The material is arranged in 15 sections (5 chapters containing 3 sections each), approximately corresponding to a course of 15 lectures. A didactic development from very basic relations at the beginning to a rather demanding level in the last sections has been striven for.

This text, as well as the lectures from which it was derived, is intended primarily for graduate students of natural sciences, for research workers in environmental sciences, and for remote sensing practitioners. The increasing degree of difficulty in the physical concepts, as well as in the mathematical methods, may cause readers of differing background knowledge to skip the one or the other section which is of less significance to their specific interest. In most cases it will be possible to continue at a subsequent chapter because in the more advanced chapters reference is only made to those of the previous sections which are related to and needed for the topic under discussion. Students of physics are expected to know most of the essentials of this text; however, it can serve to become aware which areas of physics may be relevant for use in remote sensing.

It was not the author's intention to present the newest methods of remote sensing, either to discuss the most recent details of technical developments of sensor systems and space platforms, or to describe many observational results in a phenomenological way as it is done in many texts on remote sensing. However, it is his hope to rouse the appreciation and understanding of the physical background and to create a consciousness of the importance of the interaction of electromagnetic radiation with matter for optimum use of remote sensing.

Finally, it is with great pleasure and gratitude that the author wishes to acknowledge the considerable share of a number of persons in the realization of this textbook. In the first place, I should like to thank Francis Cambou and Thuy Le Toan for the opportunity of a 6-month stay at the Centre d'Etude Spatiale des Rayonnements, Université Paul Sabatier, Toulouse, during a sabbatical semester 1982. There, for the first time, I had the opportunity to arrange and present these lectures in essentially their present form. The following colleagues at various institutions abroad helped considerably to improve the presentation and, in particular, to correct the language of different parts in this text (in alphabetical order):

J. A. Kong, Massachusetts Institute of Technology
R. W. Larson, Environmental Research Institute of Michigan
R. K. Moore, University of Kansas
P. W. Rosenkranz, Massachusetts Institute of Technology
F. T. Ulaby, University of Michigan
D. Walshaw, University of Oxford
E. R. Westwater, U.S. National Oceanic and Atmospheric Administration

Several of my colleagues at the University of Bern provided me with very valuable comments and suggestions, which helped to improve many formulations: Th. Binkert, D. A. Deranleau, R. Keese, K. Künzi, C. Mätzler. A large portion of the figures were drawn with great care, and many others were adapted for printing, by Mr. J. Brönnimann. By far the heaviest workload, however, has been carried by the secretaries of the Institute of Applied Physics, Bern, Mrs. M. Swain, Mrs. K. Gutknecht, and Mrs. B. Schweingruber, in typing, correcting and retyping the manuscript. I also wish to thank the publishing company, Springer-Verlag, for valuable advice and efficient co-operation. Last, but not least, my warmest thanks are due to my family for the patience during the time with the burden of additional work.

ERWIN SCHANDA

Contents

1 Some Basic Relations

1.1 Natural Parameters and Observables

In remote sensing of the environment we usually want to determine the state of one (or several) physical, biological or geographic quantities, e.g. the biomass of a specific crop, the quantity of a polluting gas in the atmosphere, or the extent and the state of the global snow cover at a given date. However, only in exceptional cases does the distance between the object to be investigated and the observing instruments allow a direct determination of the desired quantity. In most cases physical quantities of an object (e.g. the state of development between growth and maturity of a crop) can be characterized by some electromagnetic properties (e.g. the colour). Electromagnetic waves communicate this information to a remote sensor. If this sensor is able to receive and detect the information, i.e. if it is sensitive to the relevant part of electromagnetic waves, it can determine the electromagnetic properties of the object. However, in general, the sensor will not be able to determine directly the physical quantity (e.g. the maturity of the crop). This problem is even more obvious if the parts of the electromagnetic spectrum not visible to the human eye are involved.

Let us continue with the visible spectrum. A colour photograph is – geometrically speaking – a two-dimensional projection of a three-dimensional scene, and it presents a mixture of three primary colours, each with varying intensity (the colour combination can also be expressed by hue, saturation and intensity). Thus we have a five-dimensional multiplicity, which can contain an enormous amount of information on the geometry and light reflection properties of the observed scene.

In a more generalized sense we can speak of observables Y_j, where subscript j stands for the various "dimensions", when we treat the observed electromagnetic properties. In contrast to these observables, we may define the natural parameters by X_i when we deal with the physical properties which are to be determined by the remote sensing procedure. Table 1.1 presents, for the purpose of illustration, an arbitrary selection of both quantities.

Within this arbitrary choice evidently several of the quantities are already multi-dimensional themselves (hologram). Most of them are continuously variable, but there are discontinuous ones (crop types) present. The most important, and usually the most difficult, problem in remote sensing is to find the correct relation between the observables and the natural parameters. If this relation is not known or if it is ambiguous, one cannot achieve a correct

Table 1.1. Some examples of natural parameters and observables

Natural parameters	
Area of agricultural fields	X_1
Depth profile (3 levels) of moisture	X_2, X_3, X_4
Crop type	X_5
Biomass per unit surface area	X_6
Observables	
Fourier Components on a holographic plate	Y_1, Y_2
Frequency spectrum of scattered radiation	Y_3, Y_4, Y_5
Frequency spectrum of emitted radiation	Y_6
Degree of polarization of the reflected light	Y_7

translation of the observables into natural parameters. That is, one cannot achieve a correct and unambiguous interpretation of the remotely sensed data. In most practical situations the relation between observables and natural parameters can be very complicated, and exact solutions are unattainable or at least impracticable. However, in many special cases a clever choice of limitations of the range of validity allows algebraic simplifications; after this, solutions for the confined problem can be found that are useful and sufficiently accurate for the special applications.

One of the simplest relations is a linear one, which can serve as an approximation of the more complicated real function

$$X_1 = A_{11} Y_1 + A_{12} Y_2 + \ldots + A_{1m} Y_m$$
$$X_2 = A_{21} Y_1 + A_{22} Y_2 + \ldots + A_{2m} Y_m \tag{1.1}$$
$$\vdots$$
$$X_n = A_{n1} Y_1 + A_{n2} Y_2 + \ldots + A_{nm} Y_m .$$

Each natural parameter is assumed to depend linearly on each observable. Of course, coefficients A_{ij} will be zero if the respective X_i and Y_j are not dependent on each other. All the coefficients A_{ij} together form a matrix $\mathbb{A}$, which is a rectangular arrangement of the coefficients A_{ij} with n rows and m columns. If all elements A_{ij} of the matrix, i.e. all coefficients of Eq. (1.1) are known, one can compute all relevant natural parameters X_i from all measured observables Y_j, provided that the number of observables is sufficient to define the natural parameters. In general, however, the electromagnetic properties of media and waves provide a much easier way to compute the observables from the given natural parameters. It is also much easier experimentally to verify the observables while changing some of the natural parameters in a controlled manner. This means that the inverse presentation

$$Y = \mathbb{B} X , \tag{1.2}$$

with $\mathbb{B}$ a matrix of m rows and n columns, now defines each observable as a linear function of each natural parameter. Hence, if matrix $\mathbb{B}$ is determined either experimentally or by modelling, one can obtain the desired matrix $\mathbb{A}$ by inverting $\mathbb{B}$.

The inverse matrix $\mathbb{B}^{-1}$ (in our case we require $\mathbb{A} = \mathbb{B}^{-1}$) is defined by the product $\mathbb{B}\mathbb{B}^{-1} = \mathbb{I}$, where $\mathbb{I}$ is the identity matrix. The elements of the inverse matrix can, therefore, be explicitly computed according to the rules of the matrix calculus (e.g. Sokolnikoff and Redheffer 1958, Chap. 4).

In general, mathematically speaking, the inverse matrix $\mathbb{A}$ exists only when $\mathbb{B}$ is square $(m = n)$ and non-singular. This property implies that the determinant of the matrix $\mathbb{B}$ is not zero, because otherwise $\mathbb{A}$ will contain infinitely large elements.

The procedure for solving a remote-sensing problem like Eq. (1.2) for the unknown natural parameters X_i, when the observables Y_j have been measured, is called the inversion problem in remote sensing, or the method of indirect measurement. For the sake of simplicity we have assumed a linear relationship (1.1) which can be solved by inverting (1.2); in many cases, however, one has to deal with more complicated relations, e.g. integral equations.

If the system of linear equations like (1.2) with n unknowns X_i consists of only $m < n$ equations, the system is said to be underdetermined; there is no single unique solution. If $n = m + 1$, for any arbitrarily chosen value of X_1, respective "solutions" of $X_2 \ldots X_n$ can be found from the remaining part of the system of equations, which are, however, equally as arbitrary as X_1 and therefore, in general, do not comprise the correct solutions to the problem.

If there are more equations than unknowns $(m > n)$, the system is not necessarily overdetermined because several equations may say the same thing, i.e. there are redundant equations in the system. If, however, the equations are all independent − which means that none of them can be expressed by a linear combination of the others − then a system of m equations for $n < m$ unknowns is overdetermined and, in general, a contradiction exists between the equations.

In remote sensing problems it is often impossible to obtain the same number of observables as natural parameters. However, it is sometimes possible to define artificially an additional parameter, by which an additional equation with the same number of unknowns can be formulated without contradicting the existing ones. Or when there are more equations than unknowns, e.g. $m = n + 1$, one can solve the reduced system of n equations n times, each time by omitting another one of the m equations. The redundancy of the system can, in this case, be used to improve the accuracy of the results, because the measurements are subject to statistical and systematic errors.

For more detailed information on inversion problems in remote sensing and, in particular, on the more general inversion processes by iterative, non-linear and statistical methods, the reader is referred to comprehensive presentations like the one by Twomey (1977b) or the collection of various methods and applications edited by Deepak (1977).

In Chapter 5 we shall return to the problem of "inverting" observed values to natural parameter values for the special case of atmospheric sounding, where an integral equation is involved which will be reduced to a system of linear equations like (1.2).

The specific problems involved in inverting the properties (polarization, wavelength dependence, directional diversity) back to the material and shape properties of the source of radiation (or of scattering) by reconstructing the wavefronts of coherent and incoherent contributions are treated in various chapters of the volumes edited by Baltes (1978 and 1980) and by Boerner et al. (1985).

In the framework of our simpler treatment here, the following example may serve to illustrate an extreme situation where the large variety of the earth's surface parameters is underdetermined by the modest range of one observable: if one relied only on the albedo (the fraction of reflected light in the whole visible part of the spectrum) one could not discern between even extremely different surfaces such as grass fields, various rock types and the effect caused by thin cloud layers (Table 1.2).

Only by adding observables like colours, shapes, time (hours and seasons), solar incidence and polarization, does the determination of the kinds of object and some of their qualities become feasible.

A simple example of a situation where natural parameters are well determined by the observables is found in Maetzler et al. (1980). The microwave emission properties of sea ice were observed from an ice-breaker (Norsex

Table 1.2. Albedo of various surfaces (integral over the visible spectrum)

Surface	Percent of reflected light intensity
General albedo of the earth	
total spectrum	~35
visible spectrum	~39
Clouds (stratus) < 200 m thick	5 – 65
200 – 1000 m thick	30 – 85
Snow, fresh fallen	75 – 90
Snow, old	45 – 70
Sand, "white"	35 – 40 (increasing towards red)
Soil, light (deserts)	25 – 30 (increasing towards red)
Soil, dark (arable)	5 – 15 (increasing towards red)
Grass fields	5 – 30 (peaked at green)
Crops, green	5 – 15 (peaked at green)
Forest	5 – 10 (peaked at green)
Limestone	~36
Granite	~31
Volcano lava (Aetna)	~16
Water: sun's elevation (degrees)	
90	2
60	2.2
30	6
20	13.4
10	35.8
5	~60
<3	>90
Urban reflectance	~6 – 20

1983). These surface-based measurements can be used as a kind of standard to derive an algorithm for computing, from satellite data, the percentage of the ocean covered by young and old ice, respectively. The microwave sensors on board a satellite measure brightness temperatures averaged over huge areas containing open ocean, old and young ($<$ 1 year) ice. Figure 1.1 shows typical spectral behaviour of the brightness temperatures in the range between 5 and about 100 GHz for horizontal polarization of the electric field vector of the radiation and an incidence angle of 40 degrees.

In later sections of this text the polarization of the waves and the concept of brightness temperature will be treated in more detail. One may now take for granted that brightness temperature is proportional to the intensity of the radiation emitted by the respective surfaces, and that it can never be higher than the physical temperature of the media contributing to the received radiation. Thus sea water surface at freezing temperature ($\approx -1.8\,°C$) could, at most, exhibit a brightness temperature of ≈ 271.5 Kelvin. The emissivity, i.e. the efficiency with which a medium is able to radiate, reduces the brightness temperature to values considerably lower than the physical temperature. The emissivity characterizes a medium as a signature.

Returning to our example: suppose a microwave radiometer observes the earth's surface from a satellite at two frequencies $v_1 = 10.4$ GHz and $v_2 = 36$ GHz (about 3 and 1 cm wavelengths, respectively), and at one given point over the Arctic Ocean it measures at both wavelengths the same brightness temperature $T_1 = T_2 = 180$ K (spatially averaged over the "foot-

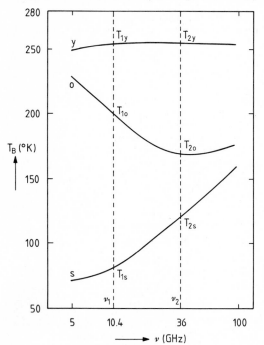

Fig. 1.1. Brightness temperatures of the open sea (s), young sea ice (y) and old sea ice (o) over the microwave range. (Maetzler et al. 1980)

print" of the antenna beam). The question to be solved is what percentage of this ocean area is covered by multi-year ice? We write Eq. (1.2) explicitly

$$T_1 = C_s T_{1s} + C_y T_{1y} + C_o T_{1o}$$
$$T_2 = C_s T_{2s} + C_y T_{2y} + C_o T_{2o} \tag{1.3}$$
$$1 = C_s + C_y + C_o ,$$

where C_s, C_y, C_o are fractions of the surface occupied by the sea, young and old ice, respectively. These fractions are the unknown natural parameters. T_{1s}, T_{1y}, T_{1o} and T_{2s}, T_{2y}, T_{2o} are the characteristic brightness temperatures of these three surface types at frequencies v_1 and v_2, respectively, and can be taken from Fig. 1.1. They constitute the coefficients of the first two linear equations in (1.3). The left sides of Eq. (1.3) are the observables. Before inverting the (3×3)-matrix, the third equation can be utilized to reduce the system to two equations for the two unknowns C_y and C_o.

$$T_1 - T_{1s} = C_y (T_{1y} - T_{1s}) + C_o (T_{1o} - T_{1s})$$
$$T_2 - T_{2s} = C_y (T_{2y} - T_{2s}) + C_o (T_{2o} - T_{2s}) . \tag{1.4}$$

This equation is the equivalent of $Y = \mathbb{B} X$ and has to be inverted to the equivalent of $X = \mathbb{A} Y$ in order to obtain the results

$$C_y = \frac{(T_1 - T_{1s})(T_{2o} - T_{2s}) - (T_2 - T_{2s})(T_{1o} - T_{1s})}{(T_{1y} - T_{1s})(T_{2o} - T_{2s}) - (T_{2y} - T_{2s})(T_{1o} - T_{1s})}$$
$$C_o = \frac{(T_2 - T_{2s})(T_{1y} - T_{1s}) - (T_1 - T_{1s})(T_{2y} - T_{2s})}{(T_{1y} - T_{1s})(T_{2o} - T_{2s}) - (T_{2y} - T_{2s})(T_{1o} - T_{1s})} . \tag{1.5}$$

Inserting the six characteristic temperatures read from Fig. 1.1 for 10.4 and 36 GHz, and using the observed brightness temperatures $T_1 = T_2 = 180$ K, Eq. (1.5) yields $C_y = 0.3$, $C_o = 0.4$, and hence $C_s = 0.3$.

From the above discussion it is clear that one needs as many observables as unknown parameters or – for the sake of reducing the effect of statistical measurement errors by redundant observations – as many observables as possible. However, the number of observables is, from a rational point of view, limited to those which are independent of each other or, in other words, limited to the number beyond which no effective increase in information is obtained. As soon as the result of one of m observations is predictable within the measurement accuracy, this measurement is totally redundant and contributes no additional information.

To collect as many independent observations as possible, one should make use of the strong wavelength dependence of the emission and scatter properties of the media. Within the large spectrum between the gamma-rays and the radiowaves which is accessible for remote sensing purposes, much information is available. Figure 1.2 is a schematic presentation of this huge part of the electromagnetic spectrum accompanied by a list, necessarily incomplete, of the types of sensors and methods used for remote sensing purposes.

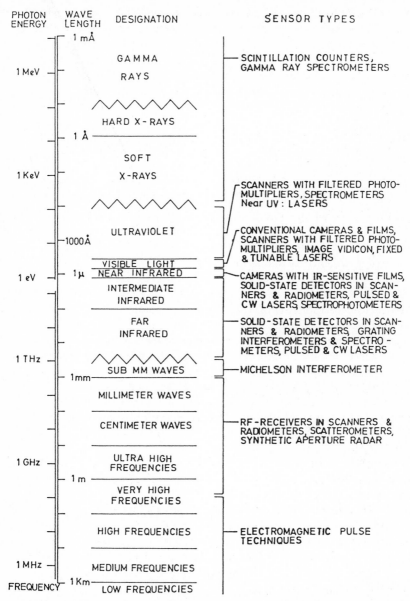

Fig. 1.2. The electromagnetic spectrum in units of wavelength, photon energy or frequency, whatever is appropriate. The various ranges of the continuous spectrum are named with their widely used designations. Different sensor types are required for the remote probing in the various ranges of the spectrum

In addition to the spectral diversity one can also make use of the different origins of the radiation. Figure 1.3 indicates three basic types of radiation which may enter the electromagnetic sensor. Radiation can be emitted by the object itself due to its material properties and physical conditions. Two examples are gamma-radiation due to radioactive substances and the thermal radiation in the infrared and microwaves due to the temperature and emissivity of the media. Second, received radiation can be caused by the diffuse scattering of natural illumination originating in a source of incoherent radiation like the sun or, more generally, daylight. Finally, an artificial coherent signal can be transmitted towards the object and a sensor, tuned for optimum reception of that signal, receives the waves after reflection at the surface of the object. In this coherent case the different contributions to the reflection by different elementary scatterers may suffer from strong interference, while with reflection of incoherent radiation (albedo), interference can occur only on a very small scale. Thus different kinds of interactions between radiation and medium participate in the radiation process; therefore different (in favourable cases: independent) types and amounts of information on the medium can be gained by utilizing different types of radiation processes.

A rough sketch of the flow of information in a complete remote sensing system may be instructive (Fig. 1.4). Between the object where the translation from the natural parameters to the observables takes place, and the sensor, which is expected to detect the observables, we have to assume some kind of disturbing interference without any relation to the object, for example atmospheric transmission properties which the sensor measures, cannot be distinguished from the radiation properties of the object. Even if the sensor properties (type, wavelength, polarization, measuring conditions etc.) are designed optimally for the object under investigation, some kind of reference

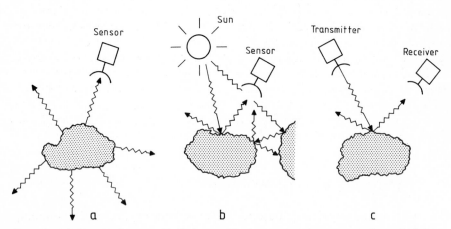

Fig. 1.3a – c. Three basic types of radiation used in electromagnetic remote sensing. **a** Emission; **b** diffuse scattering of incoherent light; **c** reflection of artificial, coherent waves

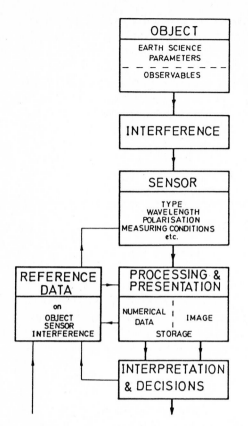

Fig. 1.4. Block diagram of a remote sensing system with optional automatic interpretation and decision. A complete system will comprise of several complementary types of sensors. The reference data will be generated by previous knowledge and by automatic adjustment

data and a priori knowledge on object, sensor and interference are needed for a correct inversion, e.g. interpretation of the measured quantities into the natural parameters of the object. One can regard the sensor as nothing more than a transformer of the received information on the observables into an image or some kind of signal to be used in data processing. In the data processing, the backwards translation from the language of observables to that of natural parameters has to be realized. Thus the relation between observables and natural parameters is the key to remote sensing. This is the reason and the motivation for treating the fundamental physics of the interaction between radiation and media in the remainder of this text.

1.2 Propagation of Electromagnetic Waves

The electromagnetic radiation utilized for remote sensing, in particular radio, infrared and visible radiation, can be described by waves; most of the important phenomena of interaction with media in remote sensing applications can be explained within this concept. There are, however, phenomena which can be described only by the particle nature of light, where the radiation is thought to be transported by photons which travel with the speed of light and whose individual energies are proportional to the respective frequencies of oscillation. In this section − and in most parts of this text − we can limit our considerations on the wave nature of electromagnetic radiation.

A wave is an oscillatory quantity (the electric and magnetic field strengths in electromagnetic waves or the gas pressure and particle speed in acoustic waves) which propagates through space. The frequency v, i.e. the number of oscillations per second, is measured in Hertz [Hz].

Let us first regard a stationary oscillation which is fixed in space. A simple non-propagating, undamped oscillation of a quantity E (we take − arbitrarily but not without intention − the electric field strength E as example) can mathematically be described by differential equation

$$\frac{d^2E}{dt^2} + \omega^2 E = 0 \ , \tag{1.6}$$

where the first term is the second-order time derivative of E and ω may be taken as a constant. Those who are familiar with electrical circuits are reminded that an electric charge on an undamped electrical circuit consisting of an inductance L and a capacitance C can cause the oscillation of an electric current I; this behaviour is described by the equation

$$\frac{d^2I}{dt^2} + \frac{I}{LC} = 0 \ .$$

This may be regarded as a special case of the equation of oscillation. Here the coefficient $1/LC$ of the second term has the same function as ω^2 in Eq. (1.6), it determines the frequency at which the circuit is resonating. A possible solution of Eq. (1.6) is

$$E = E_0 \cos \omega t \ , \tag{1.7}$$

where E_0 is the maximum value, i.e. the amplitude of the oscillation and the cosine function describes its time dependence.

This solution can be geometrically illustrated by a pointer of length E_0 which turns either clockwise or counter-clockwise about the origin of a Cartesian coordinate system with a constant speed (Fig. 1.5). The solution (1.7) is represented by the projection of the instantaneous position of a counter-clockwise rotating pointer on the x-axis. The projection on the y-axis also plotted (dashed line) in Fig. 1.5b is another possible solution, $E_0 \sin \omega t$, of Eq. (1.6). Both solutions are drawn as a function of the angle of rotation which is proportional to time t.

The resulting sine and cosine curves are identical in shape and size, they differ only by a shift of $\pi/2$ on the ωt-axis. Between $\omega t = 0$ and $\omega t = 2\pi$ one full circle, i.e. one oscillation period, is completed. If we call the duration of one period T, then the above constant ω is defined by

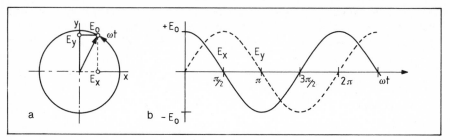

Fig. 1.5. a A pointer of length E_0 rotating at constant speed. **b** The projections of the pointer on the x- and y-axis respectively as functions of the angle of rotation (time t)

$$\omega = \frac{2\pi}{T} = 2\pi\nu \ , \tag{1.8}$$

and is called angular frequency, because the angular motion of the pointer is performed at the oscillation frequency ν, which, in its turn, is obviously related to the oscillation period as

$\nu = 1/T$.

The description of the rotating pointer by Eq. (1.7) is ambiguous, because the direction of the rotation is not determined. Therefore, it is standard notation to describe this oscillation as a motion in a complex plane with a real x-axis and an imaginary y-axis.

Thus a more complete solution of Eq. (1.6) is:

$$E = E_0(\cos\omega t + i\sin\omega t) = E_0\exp(i\omega t) \ , \tag{1.9}$$

where $i = \sqrt{-1}$ is the unit on the imaginary axis. The last part of Eq. (1.9) introduces the equivalent notation with the exponential function. The concepts of complex functions and the exponential function will be used throughout this text according to the well-established customes presented in any mathematical textbook (e.g. Sokolnikoff and Redheffer 1958 Chap. 7).

Solution (1.9) is needed when the phase of the oscillation is also considered. The dashed oscillation in Fig. 1.5 represents the sine part, which is phase shifted by $\omega t = \pi/2$ with respect to the cosine part. For the general case where the pointer does not start with $x = E_0$ at time $t = 0$, an extra constant phase term has to be added to Eq. (1.9).

If the oscillation is propagating in space, Eq. (1.6) has to be modified in order to take the spatial dependence of the oscillatory state into account. For the simplest case, the propagation into only one direction (say along the z-axis) the differential equation assumes the form

$$\frac{\partial^2 E}{\partial t^2} - c^2 \frac{\partial^2 E}{\partial z^2} = 0 \ , \tag{1.10}$$

where c is a constant, its significance and value to be determined, and ∂ stands for partial derivative. We have now two variables t and z from which the electric field strength depends. In any possible solution an oscillatory dependence along the z-direction must now appear. One of the simplest solutions of Eq. (1.10) is

$$E(z, t) = E_0 \cos(\omega t - kz) \ , \tag{1.11}$$

or the more complete version in the complex plane

$$E(z, t) = E_0 \exp i(\omega t - kz) \ , \tag{1.12}$$

where k is a constant related to the number of oscillations along the z-direction. When inserting the assumed solution (1.11) or (1.12) into (1.10) we arrive at

$$\omega^2 E = k^2 c^2 E \ ,$$

which yields the relation

$$\left| \frac{\omega}{k} \right| = c \tag{1.13}$$

between the angular frequency and the two constants c and k. If we want to know the speed of the propagating wave, we can ride with the wave on its crest starting at $t = 0$ and $z = 0$. This means that we have to move along z with the speed of the wave in order to keep the argument of the cosine constant (equal to zero) all the time, $\omega t - kz = 0$.

From this follows the wave velocity

$$\frac{z}{t} = \frac{\omega}{k} \ . \tag{1.14}$$

Comparison with Eq. (1.13) shows that the constant c in Eq. (1.10) is the wave velocity, i.e. the speed of light, $c = 3 \times 10^8 \, \text{m s}^{-1}$, when we limit our consideration to wave propagation in vacuum. This value is approximately valid also for propagation in air.

From $\omega t - kz = 0$ it follows that, while riding on the wave crest, after one time period of the oscillation T the wave must have covered a distance along z, which corresponds to a spatial period of the wave. This distance is called the wavelength: the usual symbol for it is λ. This means

$$k\lambda = \omega T = \frac{2\pi}{T} T = 2\pi \ .$$

It follows that the relation between wavenumber and wavelength is

$$k = \frac{2\pi}{\lambda} \ . \tag{1.15}$$

When Eq. (1.15) is used in (1.13) one finds

$$\frac{c}{\lambda} = \nu \ , \tag{1.16}$$

the relation between frequency, wavelength and velocity of a wave.

The differential equation (1.10) still has another solution in addition to (1.11), namely

$$E_r(z, t) = E_{0r}\cos(\omega t + kz) \ . \tag{1.17}$$

When we ride now on the wave crest starting at $t = 0$ and $z = 0$, we must move along the negative z-direction with velocity c of the wave. The solution (1.17) describes a wave moving backward. Figure 1.6 summarizes the time and space dependence as well as forward- and backward-moving waves.

For example, Eq. (1.11) describes a wave transmitted towards an object and Eq. (1.17) is the wave reflected back by this object. The absolute values of ω and k remain the same for both waves; however, the reversed sign can be attached to the wave number when the direction of propagation is reversed. From this we conclude that the wave number may be regarded as a vector. With the generalized concept of a wave vector k, the propagation of waves into any direction can be described and a more general solution of the undamped wave Eq. (1.10) becomes

$$E(r, t) = E_0 \exp i(\omega t - k \cdot r) \ , \tag{1.18}$$

where r is the vector (distance and direction) of the spatial coordinate, $k \cdot r$ the scalar or dot product, is equivalent to $kr\cos\alpha$, the product of the absolute values of k and r multiplied by the cosine of the angle between the directions of k and r.

In the case of electromagnetic waves the oscillating quantities are the electric and the magnetic fields E and H respectively. Both are vectors also and, when propagating in homogeneous isotropic media, the directions of these two field vectors are perpendicular to each other and both are perpendicular to the direction of propagation, i.e. to the wave vector k. Thus E and H propagate along k, while oscillating perpendicular to k. The direction of the oscillation of E is usually called the direction of the polarization. In

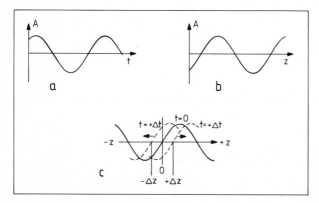

Fig. 1.6a – c. Propagating waves. Analogy between **a** time and **b** space frame; **c** forward and backward propagation

vacuum and in isotropic media the direction of the polarization is preserved. These statements on the electromagnetic fields and the propagation of waves can be exactly derived from Maxwell's equations. This will not be done here in order to avoid any overloading of this text.

However, when deriving electromagnetic waves from Maxwell's equations, one important result is the relation between the wave velocity and the material constants of the medium, where the wave is propagating, the electric permittivity ε and the magnetic permeability μ. This relation reads as

$$c^2 = \frac{1}{\varepsilon\mu} \; .$$

The values in the vacuum are (in rationalized $mksA$-units) $\varepsilon_0 = 8.85 \times 10^{-12}$ As/Vm (ampere seconds divided by volt metre) and $\mu_0 = 12.5 \times 10^{-7}$ Vs/Am (volt seconds divided by ampere metre). Most media encountered in remote sensing of the earth have permittivities of $\varepsilon = \varepsilon_r\varepsilon_0$ with the relative values (or dielectric constants) in the range $1 < \varepsilon_r < 100$, with most abundant values of ε_r close to the lower limit, while the permeabilities are usually $\mu = \mu_0$. The consequence of this property is that the wave velocity is reduced to

$$v^2 = \frac{1}{\varepsilon_r\varepsilon_0\mu_0} = \frac{c^2}{\varepsilon_r} \; , \tag{1.19}$$

where hereinafter c is the wave velocity in vacuum; therefore the valid relation in vacuum will be

$$c^2 = \frac{1}{\varepsilon_0\mu_0} \; . \tag{1.20}$$

The wave length λ in a dielectric medium is reduced correspondingly to the reduction of the wave velocity

$$\lambda = \frac{v}{\nu} = \frac{1}{\sqrt{\varepsilon_r}} \frac{c}{\nu} \; . \tag{1.21}$$

Velocity v and length λ of the wave are reduced by the factor

$$\frac{1}{\sqrt{\varepsilon_r}} = \frac{1}{n} \; , \tag{1.22}$$

where n is the index of refraction.

Up to now we have taken only a single exactly defined value for frequency ν. This hypothetical case is called a monochromatic wave. For a wave to be exactly monochromatic, the wave train must be infinitely long; moreover, a monochromatic wave could transport no information nor energy. Monochromatic waves are therefore impracticable. For a realistic description of waves one has to admit several frequencies around one center frequency ν_0, or a narrow but continuous range of frequencies called the bandwidth $\Delta\nu$ around

v_0. A change of the amplitude of the wave train (modulation) will at least need a time Δt equal to the inverse of the bandwidth. Thus information (switching off and on of the wave train) can be transmitted by a rate at most equal to Δv, the bandwidth. For example, a thermal infrared scanner is able to record the information of each pixel only if the change pixel to pixel is slower than $\Delta t = 1/\Delta v_R$ where Δv_R is the bandwidth of the recording electronics. This statement is applicable only to the flow of information from the sensor through the on-line processing to the recording. Usually a sensor has a wide spectral range of sensitivity (e.g. in the infrared). However, this kind of bandwidth is of little relevance to the limitation of the speed of recordings.

Requiring a minimum of information to be carried on a wave means one needs contributions E_i to the wave field at least at several frequencies v_i (Fig. 1.7a).

The resulting wave field at time t and at distance z can be computed by the superposition of all contributions

$$E(z, t) = \sum_i E_i \cos 2\pi v_i \left(t - \frac{z}{c} \right) . \tag{1.23}$$

From this rather academic case we can pass on to the real situation by assuming a narrow frequency region with continuously dispersed contributions to the total wave field of strengths $E(v)$ per unit bandwidth (see Fig. 1.7b). Instead of a summation over discrete values, an integral over this continuous spectrum has to be computed

$$E(z, t) = \int_0^\infty E(v) \cos \left[2\pi v \left(t - \frac{z}{c} \right) \right] dv ; \tag{1.24}$$

the unit of $E(v)$ is volt per metre per hertz.

The factor $(t - z/c)$ in the argument of the cosine signifies that at times

$$t = \frac{z}{c} \pm \frac{m}{v_0} \tag{1.25}$$

a maximum field strength occurs at the distance z, because then all spectral contributions of the signal spectrum are summed up vectorially (m is an

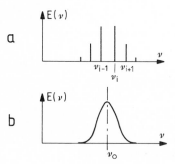

a

b

Fig. 1.7. Spectral contributions to a wave field **a** by discrete quasi-monochromatic frequency components, **b** by a continuously distributed spectrum

integer). An equivalent meaning of this factor is that a wave with the spectrum $E(v)$ launched at $z = 0$ will exhibit maximum field strengths at all distances $z = ct \pm m\lambda_0$, i.e. at all integer multiples of the mean wavelength $\lambda_0 = c/v_0$. However, with the finite width of the spectrum the wavelength is no longer defined uniquely. The consequence of this fact can be shown by performing the integration (1.24) with the simple spectrum

$$E(v) = E_0/\Delta v \quad \text{within} \quad v_0 - \frac{\Delta v}{2} \leqslant v \leqslant v_0 + \frac{\Delta v}{2}$$

and $E(v) = 0$ elsewhere.

For this case a wavepacket with the shape

$$E(z, t) = E_0 \cos\left[2\pi v_0\left(t - \frac{z}{c}\right)\right] \frac{\sin\left[2\pi\frac{\Delta v}{2}\left(t - \frac{z}{c}\right)\right]}{2\pi\frac{\Delta v}{2}\left(t - \frac{z}{c}\right)} \tag{1.26}$$

arrives at z after travel time $t = z/c$.

One easily recognizes that the infinitely small bandwidth $\Delta v \to 0$ gives the original result (1.11), while for a growing Δv the wave field packet at multiples of λ_0 becomes lower and broader. Now if $t - z/c$ deviates from the exact values given by Eq. (1.25), the resulting wave amplitude will decrease within short distances due to the destructive interference of the various spectral contributions.

Around $\left(t - \frac{z}{c}\right) \approx \frac{1}{\Delta v}$ the field strength vanishes even at distances which are multiplies of λ_0 or at times which are multiples of $\frac{1}{v_0}$. At larger values of the distance than $z \approx \frac{c}{\Delta v}$ or after longer durations than $t \approx 1/\Delta v$, the ability of the wave to sum up all spectral contributions vectorially is lost, e.g. the coherence of the wave is lost. Therefore

$$\Delta t_0 = \frac{1}{\Delta v}$$

$$\Delta z_0 = c\Delta t_0 = \frac{c}{\Delta v} \tag{1.27}$$

are called, respectively, coherence time and coherence length of a wave with the spectral width Δv. For small bandwidth ($\Delta v \ll v_0$) the relation

$$\frac{d\lambda}{dv} = \frac{d}{dv}\left(\frac{c}{v}\right) = -\frac{c}{v^2} = -\frac{\lambda^2}{c} \tag{1.28}$$

allows us to express the coherence length by the wave length $\Delta z_0 = \lambda_0^2/\Delta\lambda$.

A very coherent, artificial source of radiation (e.g. continuous wave radar) with a narrow bandwidth of, say, $\Delta v = 100$ Hz, exhibits a coherence length of $\Delta z_0 = 3000$ km, while coherence effects of a thermal radiation, even if its bandwidth is narrowed down to e.g. 1000 MHz, are extinguished after only 0.3 m. An image produced by radar or laser shows a grainy appearance (speckle) due to interferences of coherent radiation reflected by neighbouring regions of the scene.

Equation (1.24) has a more general significance as the time dependent field $E(t)$ is the Fourier transform of the spectrum $E(v)$ at some fixed point z in space, e.g. at $z = 0$.

Conversely $E(v)$ can be represented by the inverse Fourier transform of $E(t)$, thus $E(v)$ and $E(t)$ constitute a pair of functions which can be mutually expressed by a pair of Fourier transformations.

It is customary to use complex notation for a spectral function $E(\omega)$ which is defined between $\omega = -\infty$ and $\omega = +\infty$. Hence we arrive at the Fourier pair

$$E(t) = \frac{1}{2\pi} \int_{-\infty}^{+\infty} E(\omega)\exp(i\omega t)\,d\omega$$

$$E(\omega) = \int_{-\infty}^{+\infty} E(t)\exp(-i\omega t)\,dt \ . \tag{1.29}$$

The inverse transformation can be illustrated by the following example. Suppose we have a very short radar pulse, the duration of which is only a few oscillations long (say, Δt seconds) at a center frequency v_0. We ask how much bandwidth is needed for the transmitter to generate and for a receiver to detect unambiguously this pulse. The desired shape of the pulse in the time domain is given by (see Fig. 1.8 a)

$$E(t) = E_0 \cos 2\pi v_0 t \quad \text{during} \quad -\frac{\Delta t}{2} \leqslant t \leqslant +\frac{\Delta t}{2}$$

and

$$E(t) = 0$$

at other times.

The inverse Fourier transform yields the spectral distribution of the wave field

$$E(v) = E_0 \frac{\Delta t}{2} \left[\frac{\sin \pi(v_0 - v)\Delta t}{\pi(v_0 - v)\Delta t} + \frac{\sin \pi(v_0 + v)\Delta t}{\pi(v_0 + v)\Delta t} \right] , \tag{1.30}$$

which has, from a formal point of view, essentially the same character as Eq. (1.26). The spectrum has two equal contributions corresponding to the two terms in Eq. (1.30), one centered at $+v_0$ on the positive, and one at $-v_0$ on the negative frequency axis. The positive part is shown in Fig. 1.8 b.

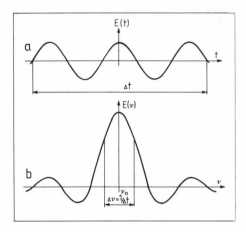

Fig. 1.8. A short ideal radar pulse, only a few oscillations long, **a** in the time domain, **b** in the frequency domain

The frequency difference Δv to the first zero's of the $\sin[\pi(v_0 - v)\Delta t]/[\pi(v_0 - v)\Delta t]$-function above and below v_0, respectively, is given by the inverse of the pulse length

$$\Delta v = \frac{1}{\Delta t} \; . \tag{1.31}$$

The width of the spectral curve between the frequencies $v_0 \pm \Delta v/2$ (drawn symmetrically in Fig. 1.8 b) contains the most important part of the pulse spectrum which is needed to reconstruct the pulse. If the bandwidth of a receiver is not wider than Δv, cutting all spectral contributions outside this frequency band, the shape of the reconstructed pulse will be strongly degraded.

The shorter the desired pulse the wider the spectrum required to realize the pulse as well as its detection. For a radar, a shorter pulse means a finer resolution of the distance and consequently more information per unit time. This higher rate of information needs more bandwidth. The formulation (1.31), or more accurately $\Delta\omega = 1/\Delta t$, is with reference to quantum physics sometimes called the uncertainty principle of communications. If the bandwidth of the receiver is not sufficient, the duration (and even the arrival time) of the pulse can only be determined with increasing uncertainty. Likewise, here as in the previous example, it is obvious that after a distance longer than $c/\Delta v$, the coherence of the signal is lost.

The concept of Fourier transformation can be utilized with another pair of variables relevant to remote sensing applications. The beam patterns of an optical objective or of a radar antenna at large distances from the objective or from the antenna (Fraunhofer zone) are given by the Fourier transforms of the spatial wave field distribution over the lens (or antenna) area. This can most easily be understood for the situation of a radar transmitter: the distribution of the electromagnetic field over the antenna surface will obviously have an effect on the shape and width of the beam within which the pulse is transmitted. Let $E_A(x, y)$ be this field distribution over the aperture area and let the angular dependence of the beam be described by the gain function $g(k_x, k_y)$ where the angles ϑ and φ of polar coordinates are implicitly contained in the transverse components of the wave vector

$$k_x = \frac{2\pi}{\lambda} \sin \vartheta \cos \varphi$$

$$k_y = \frac{2\pi}{\lambda} \sin \vartheta \sin \varphi \; .$$

The wave field radiated from an antenna is, of course, also dependent on the z-coordinate, i.e. the direction of propagation, and on the time t. As long as linear superposition of wave fields is permitted (linear regime of the material properties involved) the complete description of the electric field can be separated into factors as $E(x, y, z, t) = E(x, y)E(z)E(t)$. For our present purpose we regard only the x-y-dependence.

The two-dimensional Fourier transform is

$$g(k_x, k_y) = \iint \frac{E_A(x, y)}{E_{A\,max}} \exp\mathrm{i}[k_x x + k_y y] \, dx \, dy \; , \tag{1.32}$$

where the integration is over the surface of the antenna (or lens). A somewhat simplified example may serve as an illustration. Assume a uniformly distributed field, i.e. $E_A(x, y)/E_{A\,max} = 1$ over a rectangular aperture of the size $\Delta x \, \Delta y$, (see Fig. 1.9). The transform (1.32), i.e. the integration from $-\Delta x/2$ to $+\Delta x/2$ and from $-\Delta y/2$ to $+\Delta y/2$, yields a field strength in the far-field zone, for which the angular distribution is determined by the amplitude gain function (in units of m²)

$$g(\vartheta, \varphi) = \Delta x \, \Delta y \frac{\sin\left[\pi \dfrac{\Delta x}{\lambda} \sin \vartheta \cos \varphi\right]}{\pi \dfrac{\Delta x}{\lambda} \sin \vartheta \cos \varphi} \frac{\sin\left[\pi \dfrac{\Delta y}{\lambda} \sin \vartheta \sin \varphi\right]}{\pi \dfrac{\Delta y}{\lambda} \sin \vartheta \sin \varphi} \; . \tag{1.33}$$

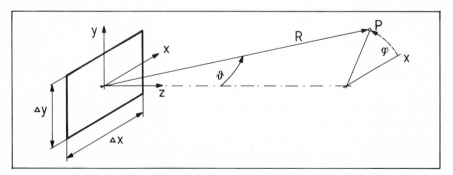

Fig. 1.9. Geometry for the determination of the gain function of a rectangular aperture

The complete formulation of the field strength E_P at a point P in the far-field zone is related to $g(\vartheta, \varphi)$ according to

$$E_P(R, \vartheta, \varphi) = \mathrm{i}\, \frac{E_{A\,\max}}{\lambda R} \exp\left(-\mathrm{i}\, \frac{2\pi}{\lambda} R \right) g(\vartheta, \varphi) \ , \tag{1.34}$$

where the dependence of amplitude and phase on the distance R are also taken into account.

Formula (1.33) shows essentially the same behaviour as the transforms (1.26) and (1.29) from a temporal to a spectral function. However, now we have a transform between a pair of two-dimensional space coordinates on the aperture and two-dimensional angular coordinates in the far field zone. The beam shape as a function of ϑ will be similar to the spectrum of the pulse (Fig. 1.8b) and the maximum gain of the wave field will be determined by the surface area of the aperture $\Delta x\, \Delta y$, the dependence on φ is only important if Δx and Δy differ strongly. To simplify Eq. (1.33), let us assume a quadratic shape of the antenna $\Delta x = \Delta y$ and its size considerably larger than the wavelength $\frac{\Delta x}{\lambda} \gg 1$: this latter condition allows us to limit our considerations to small polar angles $\vartheta \ll 1$, so $\sin \vartheta \approx \vartheta$ may be used.

Regarding a section through the beam in the x-z-plane, i.e. at $\varphi = 0$, and assuming z as the axis of beam, Eq. (1.33) is simplified to

$$g(\vartheta) \approx (\Delta x)^2 \, \frac{\sin \pi \dfrac{\Delta x}{\lambda} \vartheta}{\pi \dfrac{\Delta x}{\lambda} \vartheta} \ . \tag{1.35}$$

The larger the ratio $\Delta x / \lambda$, the smaller the angle ϑ where the first zero of the beam pattern appears. This angle is found due to

$$\frac{\Delta x}{\lambda} \vartheta = 1 \ .$$

Defining the angular resolution of an optical system (antenna) as $\Delta \vartheta$, half the width of the main beam between the first zeros of the pattern, then

$$\Delta \vartheta \approx \frac{\lambda}{\Delta x} \tag{1.36}$$

is another formulation of the "uncertainty principle". It relates the beam-width and the antenna size. If the size of Δx becomes smaller, the objects which can be resolved have to be larger.

In practice, optical lenses and many radar antennas have circular cross-sections. The Fourier transformation (1.32) then has to be formulated for circular symmetry and the result will be given in Bessel functions instead of

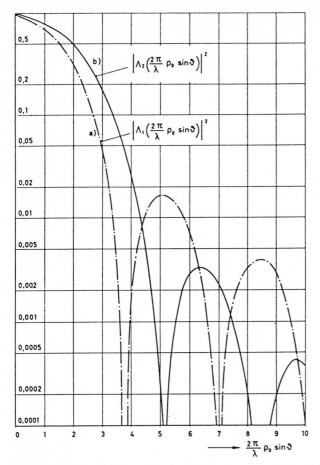

Fig. 1.10. Gain function for the intensity $|g(\vartheta)|^2$ of circular apertures of radius ρ_0 as a function of the angle ϑ. The lambda functions Λ_1 and Λ_2, related to Bessel functions of first and second order, result from the Fourier transformation of $E_A(\rho) = E_{A\,\max}$ and $E_A(\rho) = E_{A\,\max}\,[1 - (\rho/\rho_0)^2]$, respectively

the angular function in Eq. (1.33). Figure 1.10 is a presentation of the intensity gain function (proportional to $|g(\vartheta)|^2$ on a logarithmic scale for circular apertures with radius ρ_0.

Curve "a" gives the beam shape for a uniformly distributed wave field on the aperture surface $[E_A(\rho) = E_{A\max}]$, as is usually the case in optics. The angular resolution for this aperture (half the width of the main beam between the first zeros of the pattern) is approximately given by

$$\Delta\vartheta \approx \frac{3}{5}\,\frac{\lambda}{\rho_0}\ .$$

An alternate definition of the resolution, mainly used in radar techniques, using the width between the half-power-points of the beam, yields $\Delta\vartheta \approx \dfrac{\lambda}{2\rho_0}$ for case "a" in Fig. 1.10. This is the ratio of the wavelength to the diameter of the aperture.

Curve "b" is for a field distribution $E_A(\rho)$ on the aperture which is maximum at the center and decreases towards the rim of the aperture, as is usually the case for radar antennas. With this kind of field distribution on the aperture, i.e. reduced utilization of the outer parts of the antenna, the width of the resulting main beam becomes larger and the angular resolution becomes worse. Our considerations of Fourier transformation between field distribution at the aperture plane and the radiation beam are based on the assumption that the phase of the aperture field is constant over the whole aperture plane. In practice, this is usually the case or at least what is striven for.

As an example, take the case of an optical lens with a 5-cm diameter, for use at 0.5 μm wavelength on a satellite orbiting at 1000 km above the earth. The smallest objects which can be resolved according to our considerations have to be at least 10 m large.

1.3 Waves at Boundaries Between Different Media

Due to their well-defined phase information, coherent waves have the property of interfering directly. The wave amplitudes can be superimposed vectorially and, dependent on their instantaneous phase relations, they can extinguish each other or intensify to a larger resulting amplitude. For example, if a radar wave is transmitted to two neighbouring objects of equal reflectivity, the amplitudes of the individual waves scattered back to the receiver can either add up or extinguish each other, if they are in phase (difference of phase angles $\Phi = 0, \pm 2\pi, \pm 4\pi \ldots$), or out of phase ($\Phi = \pm \pi, \pm 3\pi \ldots$), respectively. In all intermediate cases the new wave amplitudes are vector sums of the two component waves with intermediate phase angles and amplitudes. Interference effects are important in all radar and some lidar applications in remote sensing.

Incoherent waves (natural light, thermal radiation), however, can only interfere by superposition of amplitudes, if the different individual scattering centers are at mutual distances closer than the coherence length, but this is, depending on the bandwidth of the receiver, usually very short. For this latter type of radiation, the intensity of the wave is the important quantity to deal with.

The intensity is given by the vector product of the electric and the magnetic field strengths; both are vectors orthogonal to each other, resulting in a new vector perpendicular to both field vectors. Consequently, in homogeneous media the vector of the intensity points in the same direction as the vector of the wave propagation k. With harmonically oscillating waves the intensity is (in units of watt m^{-2})

$$I = \tfrac{1}{2} E \times H^* \, , \tag{1.37}$$

where the $\times$ is the symbol for vector multiplication and the asterisk $*$ stands for complex conjugate. For perpendicular E and H (1.37) is numerically equal to $\tfrac{1}{2} E H^*$, if we deal with isotropic media. If the radiation is beamed within a well-defined cross-section the intensity times the cross-section yields the total power transmitted.

The ratio between electric and magnetic fields is another important quantity, the wave impedance

$$\frac{E}{H} = Z = \sqrt{\frac{\mu}{\varepsilon}} \tag{1.38}$$

in units of Ohm. In vacuum, and approximately in air, the wave impedance is $Z_0 = \sqrt{\mu_0/\varepsilon_0} = 377 \ \Omega$.

However, in a medium with $\varepsilon = \varepsilon_r \varepsilon_0$ and $\mu = \mu_0$ the wave impedance is $Z_0/\sqrt{\varepsilon_r} = Z_0/n$, reduced by the index of refraction n. The intensity within an isotropic medium is

$$I = \frac{1}{2} \frac{E^2}{Z} = n \frac{\varepsilon_0 c}{2} E^2 \tag{1.39}$$

if we use the electric field strength E in the medium. However, the field strength within the medium equals that in air only under very special conditions. We shall discuss this topic after a paragraph on the polarization of the radiation.

Let us first compare the intensities of two very different radiations, both of which playing a significant role in remote sensing. At the earth's surface we receive a solar energy flux density over the whole spectrum (solar irradiance) of approximately $1370 \ \mathrm{W \ m^{-2}}$ which is about $2 \ \mathrm{cal \ cm^{-2} \ min^{-1}}$. If this intensity were transported in a coherent wave, the amplitude of the electric field would be $E \approx 1000 \ \mathrm{V \ m^{-1}}$ (connected with a magnetic field amplitude of $H \approx 2.6 \ \mathrm{A \ m^{-1}}$). This compares with a high-power radar transmitting 1 MW into a pencil beam of a solid angle 10^{-3} steradians (units of solid angle). At a distance of 1 km the intensity is approximately $1000 \ \mathrm{W \ m^{-2}}$ which is almost the same intensity as received from the sun, most of it in the visible part of the spectrum.

During the discussion of Eq. (1.18), the solution of the wave equation, the direction of oscillation of the electric field was identified as the direction of polarization of the wave. It is well known that, through a polarizing filter, only that part of the impinging light can pass of which the polarization is parallel to the preferential orientation of the filter. The electric field is a vector which can be thought of as a vector sum composed of two orthogonal component vectors, one parallel and the other perpendicular to the preferred orientation. If the field vector is oriented 45° with respect to the preferred orientation of the filter, both the passing wave field and the field of the suppressed wave are $1/\sqrt{2}$ times the original field strength. The intensity of unpolarized light (e.g. the daylight) is divided into one half which can pass and one half which is suppressed.

If a wave is coherent, it is always polarized, and the two orthogonal components (say, in x- and y-direction, while z is the direction of propagation) can be related by their mutual phase difference Φ as

$$E_x = E_{0x}\cos(\omega t - kz)$$

$$E_y = E_{0y}\cos(\omega t - kz + \Phi) \ ,$$ (1.40)

where the total instantaneous field strength is

$$E = E_x + E_y$$

and E_{0x}, E_{0y} are the respective amplitudes. E_x and E_y have the same wavelength, but assume their respective maxima at different times t and/or distances z, if $\Phi \neq 0$ or $\neq n2\pi$. Figure 1.11 shows schematically that linear polarization can be described by (a) $\Phi = 0$, $\pm 2\pi$ ... and (b) $\Phi = \pm\pi$, $\pm 3\pi$ ... in two different orientations, while the ratio E_{0x}/E_{0y} determines the exact direction of the resulting linear polarization.

The phase differences (c) $\Phi = \dots -\dfrac{3\pi}{2}$, $+\dfrac{\pi}{2}$, $+\dfrac{5\pi}{2}$ ... and (d) $\Phi = \dots$ $-\dfrac{\pi}{2}$, $+\dfrac{3\pi}{2}$, $+\dfrac{7\pi}{2}$ ... result in clockwise and anticlockwise elliptic polarization respectively; for $E_{0x} = E_{0y}$ one obtains circularly polarized waves. Any value different from the ones above, independent of the ratio E_{0x}/E_{0y}, results in elliptic polarization. All superpositions of coherent waves at a single frequency always yield a wave with a well-defined polarization.

The case of incoherent waves (thermal radiation, daylight) is different insofar because the definition of phase becomes irrelevant [$\Phi(t)$ is a statistically varying quantity] and therefore the concept of field amplitudes is also irrelevant. Instead, we have to deal with mean square values of the field

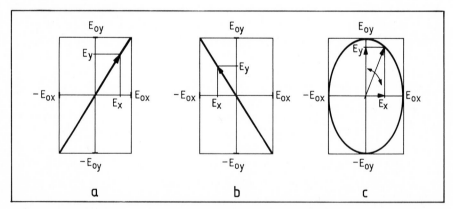

Fig. 1.11a – c. Resulting wave polarizations dependent on the phase difference between the component waves. The tip of the pointer runs along the respective *bold lines*

components $\langle E_x^2(t) \rangle$ and $\langle E_y^2(t) \rangle$ where the averages are to be taken over times larger than the reciprocal bandwidth. The sum

$$\langle E_x^2 \rangle + \langle E_y^2 \rangle = 2ZI \tag{1.41}$$

is proportional to the total intensity of the radiation. Obviously it is irrelevant for the intensity whether the wave is polarized or not. If the radiation is unpolarized, then for any choice or orthogonal x-y-coordinates we find $\langle E_x^2 \rangle = \langle E_y^2 \rangle$. If a portion of the wave intensity is polarized, then the difference

$$\langle E_x^2 \rangle - \langle E_y^2 \rangle > 0 \tag{1.42}$$

is proportional to the excess of intensity linearly polarized in the x-direction over that in y-direction, while

$$2 \langle E_x E_y \cos \Phi \rangle = 2\,\mathrm{Re}\,\langle E_x E_y^* \rangle > 0 \tag{1.43}$$

gives the excess of linear polarization in the direction of $+45°$ over that in the direction $-45°$ from the x-axis. Finally, the quantity

$$2 \langle E_x E_y \sin \Phi \rangle = 2\,\mathrm{Im}\,\langle E_x E_y^* \rangle > 0 \tag{1.44}$$

is proportional to the excess of intensity with right-hand circular polarization over that with left-hand circular polarization. For a complete description of all possible states of polarization of incoherent radiation one needs four independent parameters. These four parameters, usually called the Stokes parameters I, Q, U, V, are defined by the Eqs. (1.41) to (1.44), respectively. For the most general case, partially polarized radiation, the ratio between the intensity of polarized radiation and the total intensity is defined as the degree of polarization.

After these paragraphs on intensity, impedance and polarization of a wave we can devote ourselves to the problem of an electromagnetic wave impinging on a boundary between two different media. Figure 1.12 shows a section

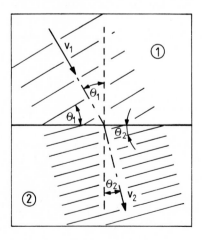

Fig. 1.12. Refractions of waves at an interface with $n_2 > n_1$ (Snellius' law)

through such a boundary region perpendicular to the plane interface and parallel to the direction of propagation of the incident wave.

Media 1 and 2 are characterized by the indices of refraction n_1 and n_2 respectively. The well-known law of Snellius tells us that the direction of wave propagation is closer to the perpendicular in the medium with the higher index of refraction according to

$$\frac{\sin \Theta_1}{\sin \Theta_2} = \frac{v_1}{v_2} = \frac{n_2}{n_1} \ . \tag{1.45}$$

$v_1 = c/n_1$ and $v_2 = c/n_2$ are the wave velocities in the two media. The relation (1.45) can be visualized very easily if we regard the phase fronts indicated by lines perpendicular to the propagation directions. The wave is slower in the medium with the higher index of refraction, therefore the distance between the wavefronts is shrunk.

Imagine Fig. 1.12 as a snapshot and, hence, the distances between neighbouring phase fronts are given by

$$\mathbf{k} \cdot \mathbf{r} = 2\pi$$

along the respective propagation directions.

The combination of Eqs. (1.15) and (1.21) shows that

$$\mathbf{k} \cdot \mathbf{r} = n\mathbf{k}_0 \cdot \mathbf{r} \ ,$$

where $\mathbf{k}_0$ is the wave vector in vacuum.

Thus the distances of the phase fronts in the two media are inversely proportional to the respective indices of refraction. There is obviously a limiting angle, Θ_{2L}, in the medium with the higher index beyond which the refracted ray cannot be directed. It will be reached when the ray in medium 1 ($n_1 < n_2$) is incident parallel to the boundary ($\Theta_1 = 90°$). It follows from Eq. (1.45) that $\sin \Theta_{2L} = n_1/n_2$.

Only a part of the wave energy incident on a boundary is, after refraction, penetrating into medium 2. Another part is reflected back into medium 1, the reflected fraction of the incident wave being also determined by the indices of refraction on both sides of the boundary. We assume again a plane boundary and for the sake of a simplified derivation we regard only the case of perpendicular incidence ($\Theta_1 = \Theta_2 = 0$).

First the boundary conditions for the electric and the magnetic fields must be taken into account (see Fig. 1.13). For field lines parallel to the plane of the interface, the electric and magnetic field strengths $E_\parallel$ and $H_\parallel$ respectively are conserved when crossing the boundary.

For field lines perpendicular to the interface plane the electric displacement $D_\perp = \varepsilon E_\perp$ and the magnetic induction $B_\perp = \mu H_\perp$ are conserved. Thus we have

$$E_{1\parallel} = E_{2\parallel} \ , \qquad H_{1\parallel} = H_{2\parallel}$$
$$\varepsilon_1 E_{1\perp} = \varepsilon_2 E_{2\perp} \ , \qquad \mu_1 H_{1\perp} = \mu_2 H_{2\perp} \ . \tag{1.46}$$

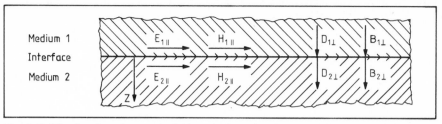

Fig. 1.13. The electric and magnetic fields at the boundary between two different media

For the wave incident perpendicularly to the interface there are no field components perpendicular to the interface, because a plane electromagnetic wave has only transverse field components. The fields at both sides of the boundary are in

Medium 1: $E_1 = E_{1f} \exp i(\omega t - k_1 z) + E_{1r} \exp i(\omega t + k_1 z)$

$H_1 = H_{1f} \exp i(\omega t - k_1 z) - H_{1r} \exp i(\omega t + k_1 z)$

Medium 2: $E_2 = E_{2f} \exp i(\omega t - k_2 z)$

$H_2 = H_{2f} \exp i(\omega t - k_2 z)$,

where the subscripts f and r stand for forward and reflected; the minus sign of the H_{1r}-term is a consequence of the relation between E and H in Maxwell's equations. E_{1f} is, of course, identical to the electric field strength carried by the original incident wave. With the origin of the coordinate z at the interface and arbitrarily making $t = 0$, the boundary conditions [i.e. the first line of, Eq. (11.46)], are

$$E_{1f} + E_{1r} = E_{2f} \tag{1.47a}$$

$$H_{1f} - H_{1r} = H_{2f} . \tag{1.47b}$$

From Eq. (1.38) and the discussion thereafter, we know that $H = nE/Z_0$. When this is used in Eq. (1.47b) for the magnetic fields, we obtain for it

$$n_1 (E_{1f} - E_{1r}) = n_2 E_{2f} . \tag{1.47c}$$

Combining (1.47a) and (1.47c) we obtain two equations

$$\frac{E_{2f}}{E_{1f}} - \frac{E_{1r}}{E_{1f}} = 1$$

$$\tag{1.48}$$

$$\frac{n_2 E_{2f}}{n_1 E_{1f}} + \frac{E_{1r}}{E_{1f}} = 1$$

for the reflection coefficient r_E of the electric wave field in medium 1 caused by the boundary and for the transmission coefficient t_E of the electric wave field from medium 1 into medium 2 across the boundary. The solutions of Eq. (1.48) are

$$r_E = \frac{E_{1r}}{E_{1f}} = \frac{n_1 - n_2}{n_1 + n_2}$$

$$t_E = \frac{E_{2f}}{E_{1f}} = \frac{2n_1}{n_1 + n_2} \ .$$

(1.49)

The corresponding coefficients for the magnetic field are

$$r_M = -r_E \ , \qquad t_M = \frac{n_2}{n_1} t_E \ .$$

(1.50)

These results can only be proven, in a simple way, if they can be used to describe the reflection and transmission coefficients of the wave intensity (also called reflectivity and transmissivity, respectively) because the intensity, being proportional to the wave energy, is conserved. Media 1 and 2 are homogeneous and isotropic except at the boundary. Therefore, the intensity can be represented by $I = \frac{1}{2} EH^*$ and reflectivity and transmissivity can also be computed accordingly.

$$r_I = r_E r_M^* = -\left[\left(1 - \frac{n_2}{n_1}\right) \Big/ \left(1 + \frac{n_2}{n_1}\right)\right]^2$$

$$t_I = t_E t_M^* = 4 \frac{n_2}{n_1} \Big/ \left(1 + \frac{n_2}{n_1}\right)^2 \ .$$

(1.51)

The negative sign of r_I indicates the reversal of the energy at the reflection. The proof $|t_I| + |r_I| = 1$ can easily be made.

If the wave does not arrive perpendicularly at the boundary plane, $\Theta \neq 0$, one has to distinguish between the different polarization components of the electric field vector (the magnetic field polarization does not matter as long as both media have the same magnetic permeability, and we have assumed even $\mu_1 = \mu_2 = \mu_0$). If the media exhibit losses, one has to take into account that the permittivities become complex quantities $\varepsilon = \varepsilon' - i\varepsilon''$ and so do the indices of refraction. The complete equations for the reflectivities of the most general case of a plane interface are due to Fresnel (Born and Wolf 1964). With the simplifying assumptions that medium 1 is vacuum (valid for air with good approximation), i.e. $n_1 = 1$, and medium 2 is a lossy dielectric for which we express the index of refraction by the complex dielectric constant $n_2 \equiv n = \sqrt{\varepsilon_r' - i\varepsilon_r''}$, the Fresnel equations read

$$r_{Ih}(\Theta) = \frac{(p - \cos \Theta)^2 + q^2}{(p + \cos \Theta)^2 + q^2}$$

$$r_{Iv}(\Theta) = \frac{(\varepsilon_r' \cos \Theta - p)^2 + (\varepsilon_r'' \cos \Theta - q)^2}{(\varepsilon_r' \cos \Theta + p)^2 + (\varepsilon_r'' \cos \Theta + q)^2}$$

(1.52)

with the abbreviations

$$p = \frac{1}{\sqrt{2}} \{[(\varepsilon_r' - \sin^2\Theta)^2 + \varepsilon_r''^2]^{1/2} + [\varepsilon_r' - \sin^2\Theta]\}^{1/2}$$

$$q = \frac{1}{\sqrt{2}} \{[(\varepsilon_r' - \sin^2\Theta)^2 + \varepsilon_r''^2]^{1/2} - [\varepsilon_r' - \sin^2\Theta]\}^{1/2} .$$

(1.52)

The subscripts h and v are taken for horizontal and vertical polarization respectively, and Θ without any subscript has the meaning of Θ_1 from now on.

The polarizations are to be understood as linear, and "horizontal" means that the electric field vector is oriented parallel to the interface, while "vertical" polarization means that a wave arrives at the interface with a vertical component ($\sin\Theta$ times the total electric field vector), but a horizontal component also exists ($\cos\Theta$ times the total field).

Assuming both media lossless ($\varepsilon'' = 0$) the Eqs. (1.52) are simplified to

$$r_{Ih}(\Theta) = \left\{ \frac{[n^2 - \sin^2\Theta]^{1/2} - \cos\Theta}{[n^2 - \sin^2\Theta]^{1/2} + \cos\Theta} \right\}^2$$

$$r_{Iv}(\Theta) = \left\{ \frac{n^2\cos\Theta - [n^2 - \sin^2\Theta]^{1/2}}{n^2\cos\Theta + [n^2 - \sin^2\Theta]^{1/2}} \right\}^2 .$$

(1.53)

From this formulation, which is a reasonable approximation for many flat boundaries of real materials, we can recognize that the numerator of r_{Iv} vanishes at one specific incidence angle Θ_B fulfilling the relation $\tan\Theta_B = n$. This angle is called the Brewster angle. Vertically polarized radiation impinging on a flat surface at the Brewster angle completely penetrates into the medium; in other terms unpolarized radiation incident under this angle is partly reflected and this reflected radiation is horizontally polarized.

Figure 1.14 is a presentation of the amplitude reflection coefficient as a function of the incidence angle on a water surface.

The reflection coefficient of the wave amplitude rather than that of the intensity ($|r_E| = \sqrt{r_I}$) is drawn in this figure in order to illustrate more distinctly the difference between optical and microwaves as well as the Brewster effect. The permittivities of water at both wavelengths have been taken as lossless, which is quite well fulfilled in the optical range, but is a crude approximation at microwaves.

This last fact leads us to another important topic. A wave penetrating into a medium may undergo attenuation if the medium is able to absorb or to scatter the wave. As a consequence, the penetration depth of the wave is limited. The ability of a medium to absorb or to scatter a wave is strongly dependent on the wavelength of the radiation, and there are many different mechanisms on the molecular or on a macroscopic scale causing the attenuation. Several of these mechanisms will be discussed in the next few chapters of this text.

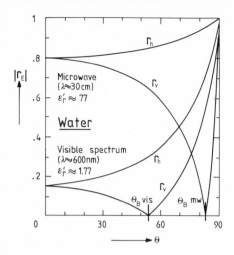

Fig. 1.14. Reflection coefficient of water as a function of the incidence angle Θ at microwaves and in the visible range for horizontal and vertical polarization

For the present purpose it is sufficient to treat the attenuating medium as a homogeneous lossy dielectric. Describing the wave propagation in the complex notation introduced by Eq. (1.12), a lossy dielectric is characterized by a complex relative permittivity $\varepsilon_r = \varepsilon_r' - i\,\varepsilon_r''$, where the real part corresponds to the usual lossless dielectric constant, while the losses are expressed by the imaginary part ε_r''. Using this complex relative permittivity in the Eq. (1.12) we first have to compute the wave number

$$k = \frac{\omega}{v} = \omega\sqrt{\varepsilon_0\mu_0}\sqrt{\varepsilon_r' - i\varepsilon_r''} = k' - i k'' = k_0(n' - i n'') \;, \tag{1.54}$$

which, along with the index of refraction, also becomes a complex quantity.

The equation of the propagating wave field now becomes, for vertical incidence

$$E = E_0 \exp i(\omega t - k' z)\exp(-k'' z) \;, \tag{1.55}$$

where z is the direction of the wave propagation perpendicular to the surface ($z = 0$ at the surface, see Fig. 1.15).

Equation (1.55) describes a wave attenuated while propagating along the z-direction with k'' being the attenuation constant. If the wave has propagated a distance z into the medium so that $k'' z = 1$, the wave amplitude is attenuated to the $1/e$ fold of the original value E_0. This distance,

$$d_A = \frac{1}{k''} \;, \tag{1.56}$$

is called the penetration depth of the wave amplitude. In many cases the penetration depth of the intensity is the important quantity. Due to Eqs. (1.37) and (1.38) the intensity is given by

$$I = \frac{EE^*}{2\sqrt{\mu/\varepsilon}} \;. \tag{1.57}$$

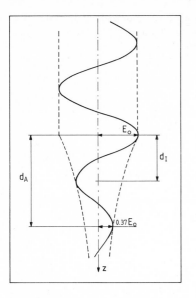

Fig. 1.15. A wave penetrating in a lossy medium

Using Eq. (1.55) in (1.57), we find that the intensity is decreasing as

$$I(z) = I_0 \exp(-2k''z) \, , \tag{1.58}$$

where I_0 is the intensity at $z = 0$. Defining the penetration depth of the intensity d_I according to the $1/e$-fold distance we arrive at

$$d_I = \frac{1}{2k''} = \frac{d_A}{2} \, . \tag{1.59}$$

The penetration depth of the intensity is half the penetration depth of the wave amplitude.

For many natural media one finds $\varepsilon_r'' \ll \varepsilon_r'$. Therefore the complex wave number may be approximated as

$$k' - i k'' \approx \frac{2\pi}{\lambda_m} \left(1 - i \frac{\varepsilon_r''}{2\varepsilon_r'}\right) \, , \tag{1.60}$$

where $\lambda_m = \lambda_0/\sqrt{\varepsilon_r'}$. The penetration depth of the intensity then becomes approximately

$$d_I \approx \frac{1}{k'} \frac{\varepsilon_r'}{\varepsilon_r''} \tag{1.61}$$

with $k' \approx \dfrac{2\pi}{\lambda_m} = \dfrac{2\pi\sqrt{\varepsilon_r'}}{\lambda_0}$.

Water exhibits a strongly frequency-dependent permittivity in the microwave range. There, the penetration depth as a function of frequency changes not only due to the changing free-space wavelength $\lambda_0 = c/v$, but, equally important, due to the changing values of ε_r'' and to changing ε_r' which in its turn affects λ_m.

Table 1.3 gives the approximate values of the permittivities and the resulting wavelengths and penetration depths of wave intensities at free-space wavelengths of 30 and 3 cm in water and moist loam. The values taken for ε_r' and ε_r'' are not according to accurate experimental verifications, and therefore the computed λ_m and d_I values are also very approximate results, only intended to illustrate the effect of the permittivity on the penetration depth. The penetration depth of water, when derived more accurately from the dielectric properties in this wavelength range, is discussed once more in Chapter 3.3.

Table 1.3. Effect of permittivity and wavelength on the penetration depth

	$v = 10^9$ Hz ($\lambda_0 = 30$ cm)				$v = 10^{10}$ Hz ($\lambda_0 = 3$ cm)			
	ε_r'	ε_r''	λ_m [cm]	d_I [cm]	ε_r'	ε_r''	λ_m [cm]	d_I [cm]
Water	80	5	3.4	8	60	30	0.4	0.15
Moist loam	10	1	9.5	15	10	5	1	0.3

Most of the material covered in Chapter 1.2 and 1.3 is treated very thoroughly and comprehensively in many standard text books on electromagnetic waves, radio waves and optics like the ones by Stratton (1941) and Born and Wolf (1964), with detailed reasoning and exact derivations.

2 Spectral Lines of Atmospheric Gases

2.1 Resonant Frequencies of Molecules

The density of the molecules in a gas at atmospheric pressure is at most one thousandth of the density in a liquid or solid material. The distances between the molecules are typically ten times the size of the molecules or more, therefore the mutual perturbation is weak. Any oscillation of a molecule, e.g. a vibration, can go on undisturbed on average during many periods before it becomes interrupted, e.g. due to the collision with a neighbouring molecule. The dynamic behaviour of the ensemble of molecules is usually influenced by incident electromagnetic waves. However, due to the weak mutual disturbance between molecules, the interaction of an individual molecule and a photon may essentially be regarded as independent of the behaviour of other molecules. The total radiation effect of the gas volume is, therefore, simply the sum of all individual contributions.

The dynamic properties of isolated atoms and molecules (e.g. their rotations, their internal motions such as vibrations and the electron orbits) can only assume certain discrete values (states) and can no longer undergo continuous changes between these states. The quantities of energy involved in the discrete elementary energy changes are so small that one has to deal with their smallest unit, the energy quantum. Thus, the dynamic behaviour of single atoms and molecules is quantized to discrete energy levels. An interaction of a molecule with light (or with any kind of electromagnetic radiation), which causes a change of the energetic state of the molecule, can take place only if the wavelength of the radiation has a very well-defined value characteristic of that molecule and of the respective pair of energy values between which the change is to take place. These facts can be expressed in quantum mechanical terms by the simple relation

$$W_n - W_m = h\nu \tag{2.1}$$

between the difference of energies W_n and W_m of an energetically higher state n and a lower one m and the frequency ν of the radiation. The factor of proportionality, the Planck constant h, has the value of about 6.6×10^{-34} W s^2 or (in units most popular in quantum mechanics) about 4.1×10^{-15} eV s.

Thus the radiation has to resonate with the energy gap, if it is to cause the molecule to make a "jump" of its total energy across this gap. If the molecule changes from a state of lower energy to one of higher energy, it needs to absorb an energy quantum out of the radiation field. This quantum of radiation is the photon. If, on the other hand, the molecule changes its energy from the upper of two levels to the lower one, the energy corresponding to this change is liberated by emitting a photon. The frequency v of the radiation has the same value in both cases, the absorption and the emission, and is given by Eq. (2.1).

A molecule has a great number of energy levels available, due to the large number of different modes of motion (rotation, vibration, electron orbits and spins), and due to the different states (ground state, excited states) within the different modes. The ability of a volume of gas to emit or absorb the radiation, corresponding to the difference between two given energy levels, depends on the number of molecules occupying the states of the higher or the lower energy respectively. By thermodynamic arguments, the ratio of these numbers per unit volume can be found to be

$$\frac{N_n}{N_m} = \exp\left[-\frac{W_n - W_m}{kT}\right].$$ (2.2)

Equation (2.2) is the Boltzmann distribution of the relative numbers of molecules over the different states. T is the absolute temperature in Kelvin (K) and

$$k = 1.38 \times 10^{-23}\,\text{Ws K}^{-1} = 8.6 \times 10^{-5}\,\text{eV K}^{-1}$$

is Boltzmann's constant. Figure 2.1 indicates schematically the number densities N_n and N_m corresponding to the upper and lower energy levels W_n and W_m respectively. The effect of a change of temperature results in a different steepness of the exponential curve connecting the respective numbers. Within an (assumed) pure two-level system, obviously $N_m + N_n = N'_m + N'_n$, the total number of molecules has to remain conserved. In normal thermodynamic conditions the upper energy state is always occupied by a smaller number of molecules than the lower state. This becomes apparent in Fig. 2.1.

If radiation of a frequency matching the difference of two energy levels is impinging on the volume of gas, an interaction can occur along three different channels:

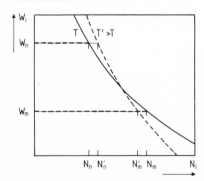

Fig. 2.1. The numbers of molecules at two energy levels W_n and W_m according to the Boltzmann distribution (2.2) for two temperatures T and $T' > T$

– The absorption process, the rate of which is proportional to the number of molecules per unit volume in the lower state N_m, to the energy density of the radiation U_ν per unit of frequency interval at the frequency ν and to a coefficient B_{mn} for absorption.

– Emission of radiation can be stimulated by the incident radiation, the probability of this process is proportional to the number density of molecules in the upper energy state N_n, to the energy density of the incident radiation U_ν and to a coefficient B_{nm} for stimulated emission.

– Finally, emission of radiation can occur spontaneously following statistical laws, only due to the fact that by the thermal agitation not only the lowest energy state (ground state) of e.g. the rotational modes, but also higher (excited) states are occupied according to Eq. (2.2). The probability of spontaneous emission is proportional only to the number density of molecules in the upper state N_n and a coefficient A_{nm} for spontaneous emission.

We can now note an equation for the dynamic equilibrium between absorption and emission:

$$N_m B_{mn} U_\nu = N_n B_{nm} U_\nu + N_n A_{nm} \, , \qquad (2.3)$$

which is a necessary condition for thermodynamic equilibrium between the electromagnetic radiation present in a given volume and the thermal agitation of the molecules in the same volume. The coefficients B_{mn}, B_{nm}, A_{nm} are the factors of proportionality and they represent the probabilities of a molecule undergoing one of these processes per unit time. They are called the Einstein coefficients for absorption, stimulated and spontaneous emission respectively. For example, the latter can be related to the natural life-time of a given excited state n by

$$\Delta t_N = 1/ \sum_m A_{nm} \qquad (2.4)$$

as the inverse of the sum of all probabilities for spontaneous change from state n to any other state. For transitions between the electronic states, interacting with visible and ultraviolet light, the natural life time is typically in the range between 10^{-9} s and 10^{-6} s. Combining Eqs. (2.1), (2.2), and (2.3) we can calculate the equilibrium energy density of the radiation U_ν per unit of frequency interval at the frequency of a transition $m - n$ of a given type of molecule for which the Einstein coefficients are known

$$U_\nu = \frac{A_{nm}}{B_{mn} \exp[h\nu/kT] - B_{nm}} \, . \qquad (2.5)$$

In order also to maintain the equilibrium for infinitely high temperature ($T \to \infty$), U_ν also has to become infinite. Accordingly, from Eq. (2.5) follows

$$B_{mn} = B_{nm} \, . \qquad (2.6)$$

From a formal point of view $T \to \infty$ is equivalent to $\nu \to 0$. However, for low radiation frequencies, such as $h\nu/kT \ll 1$, the classical Rayleigh-Jeans law gives

$$U_v = \frac{8\,\pi v^2}{c^3}\,kT\ ,$$
(2.7)

which, by comparison with Eq. (2.5), yields

$$\frac{A_{nm}}{B_{nm}} = \frac{8\,\pi h v^3}{c^3}\ .$$
(2.8)

We now assume for a moment a dense distribution of different molecules, each with many different energy levels so that the resulting spectral lines are so closely packed over a wide frequency range as to form a continuous radiation spectrum. Consequently, the frequency of oscillation is no longer fixed to a few discrete values, but may be regarded as a continuous variable. With the knowledge of the general relations between the Einstein coefficients from Eqs. (2.6) and (2.8), the famous Planck's law of thermal radiation for a blackbody can be derived

$$U_v = \frac{8\,\pi h v^3}{c^3}\left[\exp\left(\frac{hv}{kT}\right) - 1\right]^{-1}\ .$$
(2.9)

This is the radiation energy per unit of volume and per unit of frequency bandwidth, i.e. Joule m^{-3} Hz^{-1}.

Figure 2.2 presents the thermal radiation spectrum between the microwave ($\lambda = 3$ cm) and the ultraviolet ($\lambda = 30$ nm) regions for typical temperatures of the earth's surface, the cosmic background and for solar radiation reflected at the earth.

The blackbody spectral radiance $S_b(v, T)$ is presented on the ordinate axis. This quantity gives the radiant intensity per unit of surface, per unit of bandwidth and per unit of solid angle, instead of the spectral energy density. It

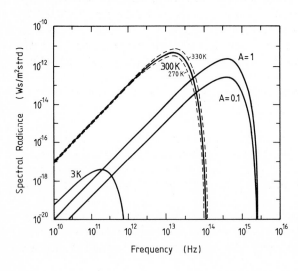

Fig. 2.2. The spectral radiance of a black body according to the Planck function for typical terrestrial temperatures (270 K, 300 K, 330 K), the cosmic background radiation (3 K) and the solar irradiance reflected from the earth's surface with two values of the albedo $A = 1$ and $A = 0.1$

differs from the latter by a factor $c/4\pi$. The radiance is particularly useful when dealing with radiating surfaces and with the transport of radiation. We shall use this concept extensively in Chapter 5 of this text.

Figure 2.2 shows that the maximum radiance is proportional to the third power of the temperature. This is a consequence of Wien's displacement law, which shifts the frequency of the radiance maximum proportional to the temperature. Thus the spectral range of maximum radiance is determined by the exponential function, the maximum value itself, however, is due to the factor v^3 in Eq. (2.9). For the computed solar irradiance reflected from the earth's surface an effective surface temperature of the sun of 5770 K has been assumed. A purely geometric "dilution" factor is $\pi R_s^2/D_{SE}^2 \approx 0.685 \times 10^{-4}$, where R_s is the radius of the solar disc and D_{SE} is the sun-earth distance, the sun is assumed to radiate as a perfect Lambertian surface and for the earth's surface two values of the albedo, $A = 1$ and $A = 0.1$, are taken for a viewing direction along the nadir at local noon.

Obviously, at wavelengths shorter than the region of intersection (at visible and near infrared wavelengths) the reflected solar radiation is observed when looking down at the earth's surface, while for longer wavelengths the thermal radiation of the earth's surface itself dominates the radiance. The wavelength range of approximately 10 to 15 µm ($v \approx 2$ to 3×10^{13} Hz), corresponding to the maximum radiance for typical terrestrial temperatures, is usually called "thermal infrared" by the remote sensing community.

More exact and detailed explanations of the above concept and the derivation of the Planck radiation formula can be found in standard text books on quantum electronics and optics such as, e.g. Haken (1981).

Let us return now to the description of the dynamic behaviour of isolated molecules. We can discern various types of motion relevant for the interaction with radiation. The electrons are spinning around axes along discrete directions, while moving on orbits around the atomic nuclei and between the atoms of a molecule. The atoms in a molecule may vibrate along or perpendicular to the bonding axis or in any combination of these directions. The whole molecule may rotate about any axis. As we shall see later in this chapter, even translational motions have an effect on the radiation emitted or absorbed by a molecule.

The total energy of a molecule is the sum of all these contributions

$$W_t = W_e + W_v + W_r + \ldots \, .$$

The quantities decisive for the interaction with radiation are the energy differences between neighbouring levels of the respective types of motions.

Table 2.1 gives the typical ranges of energies and corresponding radiation frequencies for transitions between different electronic, vibrational and rotational states.

Each of these types of transition covers a wide range of energies and corresponding spectral ranges for the large variety of species so that there is even some spectral overlap. Taking only the composition of gases in the earth's

Table 2.1. Typical energy differences and spectral ranges in electronic, vibrational and rotational transitions

Transition	ΔW [eV]	$\nu = \Delta W/h$ [Hz]	Spectral range
Electronic	10	2.4×10^{15}	Ultraviolet and visible
Vibration	10^{-1}	2.4×10^{13}	Infrared
Rotation	10^{-3}	2.4×10^{11}	Millimetre waves

atmosphere, one finds the whole spectrum between millimetric waves and ultraviolet covered by spectral lines.

We can treat the dynamic behaviour of molecules in the context of these lectures in only a very approximate way. We first regard the case of a molecule rotating about an axis orthogonal to its bond axis. For simplicity we take a diatomic molecule with masses m_1 and m_2 of the atoms at an equilibrium distance of R_0 (Fig. 2.3a). From classical mechanics we know that the molecular moment of inertia is

$$I = \frac{m_1 m_2}{m_1 + m_2} R_0^2 \ ,$$

where the atomic masses are assumed to be concentrated in volumes very small compared to their mutual distance. The angular momentum is $L = \omega I$, where ω is the angular frequency. The energy of this rotating dumb-bell model is

$$W_r = \frac{1}{2} I \omega^2 = \frac{L^2}{2I} \ . \tag{2.10}$$

Quantum mechanics teaches that the angular momentum is quantized according to

$$L = \sqrt{J(J+1)}\, \hbar \ , \tag{2.11}$$

where $\hbar = h/2\pi$ Planck's constant divided by 2π and J is called the angular momentum quantum number of a molecule, it can attain any positive integer $0,1,2,3\ldots$ leading to $L = 0, \sqrt{2}, \sqrt{6}, \sqrt{12},\ldots$ Therefore, the energy of a quantized rotating molecule becomes

$$W_r = \frac{J(J+1)}{2I} \hbar^2 \equiv BhJ(J+1) \tag{2.12}$$

instead of Eq. (2.10). $B = h/8\pi^2 I$ is the rotational constant, i.e. a factor characterizing each molecule because it contains its moment of inertia: for polyatomic non-linear molecules B assumes different values in the different directions of orthogonal coordinates. For high rotational states J the distance R_0 between the atoms is slightly stretched, leading to an increase of the moment of inertia I and a corresponding reduction in the value of W_r.

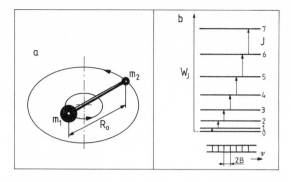

Fig. 2.3. a Model of a diatomic molecule rotating perpendicularly to its body axis. **b** Schematic presentation of the energy levels of a rigid rotator

Figure 2.3 shows a diatomic molecule rotating about its common point of gravity orthogonally to its body axis.

The increasing rotational energy with increasing quantum number J is indicated qualitatively. Quantum mechanically allowed transitions are between neighbouring states, i.e. the rotational quantum number may change by one unit $\Delta J = \pm 1$. Such a change can be caused, for instance, by the absorption of a photon, the energy ($h\nu$) of which has to match the energy difference between the two rotational states, say J and $J+1$, of the molecule according to Eq. (2.1). From Eq. (2.12), it can easily be proven that ΔW_r increases with increasing J in steps of $2, 4, 6 \ldots$ units of hB for the transitions $0-1, 1-2, 2-3 \ldots$ respectively. Of course, in order to interact with electromagnetic radiation, the molecule has to present an electric dipole moment. This dipole moment – measured in units of Debye (1 Debye $= 3.3 \times 10^{-30}$ A s m) – can be created by an asymmetric distribution of the electric charges between the atoms in the molecule.

For the case of absorption of radiation the frequency needed in order to cause this transition is given by

$$\nu_r = \frac{W_{J+1} - W_J}{h} = 2B(J+1) \; . \tag{2.13}$$

In case a photon is emitted by the rotating molecule due to a jump from state $J+1$ down to J, the same amount of rotational energy is liberated and transferred to radiation. Thus the frequency of the emitted radiation is given just as well by Eq. (2.13).

Table 2.2 gives the approximate transition frequencies for a few simple molecules occurring in the earth's atmosphere as minor constituents. We see that the ranges of transition frequencies are inversely proportional to the moments of inertia, which in their turn are determined by the mass distribution within the molecules. The distances R_0 between the atoms in a molecule are very similar in most of the diatomic molecules. In laboratory spectroscopy one can determine this distance very accurately from $I = R_1^2 m_1 + R_2^2 m_2$, where R_1 and R_2 are the distances of the two atoms (masses m_1 and m_2) from the common centre of gravity. For this purpose one has to measure the transition

Table 2.2. Approximate values of molecular parameters and rotational transition frequencies of a few diatomic molecules. (Gordy and Cook 1970, Poynter and Pickett 1984)

Molecule	R_0 $(10^{-9}\,m)$	Moment of inertia $(10^{-46}\,kg\,m^{-2})$	Dipole moment (Debye)	Transition $J \leftrightarrow J+1$	Frequency $(10^9\,Hz)$
CO	0.113	1.45	0.11	0−1	115
				1−2	231
				2−3	346
NO	0.115	1.65	0.15	1/2−3/2	151
				3/2−5/2	251
HCl	0.128	0.27	1.08	0−1	626
				1−2	1251
				2−3	1876
ClO	0.156	4.51	1.24	1/2−3/2	56
				3/2−5/2	93

frequency very accurately, and of course one has to know the atomic masses and the quantum numbers involved in the transition. The effective quantum numbers are not always integers, because in several molecules there exists a coupling of electronic angular momentum (either from the orbit or the spin of an electron) to the angular momentum of the molecular rotation in such a way as to cause half-integer rotational quantum number 1/2, 3/2, 5/2 ... Two examples are given in Table 2.2 and the effect on the transition frequencies follows from Eq. (2.13).

The electric dipole moment depends on the charge distribution in the molecular bond, and it can vary between molecules from zero (symmetric molecules as O_2) to several units of Debye. The dipole moment has no effect in a first approximation on the transition frequency; however, it determines the intensity of absorption and emission. Other details as, e.g. the hyperfine splitting of the rotational levels (as in ClO) due to interactions of the electric or magnetic molecular fields with nuclear moments, are not considered within this discussion.

Most of the molecules encountered in nature are composed of more than two atoms. Due to the complicated configurations and interactions between various internal angular momenta, the spectroscopic treatment of the rotational behaviour becomes increasingly complicated and would exceed by far the goals of this text. The interested reader is therefore referred to specific textbooks such as the ones by Howarth (1973), Gordy and Cook (1970), and Townes and Schawlow (1955).

We shall now regard, in an elementary way, some spectroscopic aspects of the vibration of a diatomic molecule. The equilibrium between the forces of attraction and repulsion between the two atoms of masses m_1 and m_2 is established at a distance R_0. A change of this distance causes a force which tends to restore this equilibrium. As indicated in Fig. 2.4a, the action of this force can be visualized by a spring with a characteristic restoring force constant

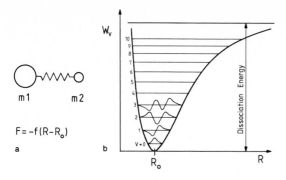

Fig. 2.4a, b. Vibration of a diatomic molecule. **a** A spring between the two atomic masses as
a model; **b** the Morse potential
curve and some vibrational
modes corresponding to the
different vibrational energy
levels

(spring constant) f (von Hippel 1954, Bingel 1967). Distances R different from the equilibrium distance R_0 between the two point-like masses cause a force

$$F = -f(R - R_0) \, . \tag{2.14}$$

In fact the deviation of the distance from the equilibrium R_0 gives rise to an increase of the potential energy. The vibrational motion of a diatomic molecule is governed by a potential function usually described by the Morse function (Fig. 2.4b)

$$V(R) = D\{1 - \exp[-\zeta(R - R_0)]\} \, , \tag{2.15}$$

where D is the dissociation energy which is needed to disrupt the molecule ($R \to \infty$) and ζ is a constant. The Morse potential curve is drawn schematically as the envelope of the horizontal strokes representing radial extent and energy levels of the various vibrational states $v = 0, 1, 2, \ldots$

The simple classical treatment of harmonic vibration with the restoring force (2.14) and the assumption of small amplitudes $R - R_0$ already leads to a useful formula for the angular frequency of the vibration

$$\omega_v = \sqrt{\frac{f(m_1 + m_2)}{m_1 m_2}} \, . \tag{2.16}$$

Quantum-mechanically the possible energy states of the harmonic vibration are given by

$$W_v = (v + \tfrac{1}{2}) \hbar \omega_v \, , \tag{2.17}$$

with the vibrational quantum number $v = 0, 1, 2, \ldots$ (e.g. Herzberg 1945, p. 77). Obviously the quantum-mechanically allowed transitions between neighbouring states are related to interactions with radiation at the vibrational frequencies ω_v according to Eq. (2.16) derived from classical arguments.

Two very different diatomic molecules HCl and CO may serve to illustrate the order magnitudes of the quantities involved. The atomic masses are $m_\mathrm{H} \approx 1.6 \times 10^{-27}$ kg, $m_\mathrm{Cl} = 58.8 \times 10^{-27}$ kg, $m_c = 20.0 \times 10^{-27}$ kg, $m_0 = 26.5 \times 10^{-27}$ kg. The total masses of the two molecules are similar, but the reduced masses $\bar{m} = m_1 m_2 / (m_1 + m_2)$, as needed in (2.16),

$$\bar{m}_{HCl} \approx 1.62 \times 10^{-27}\,kg \quad\text{and}\quad \bar{m}_{CO} = 11.4 \times 10^{-27}\,kg$$

are different by a factor of 7.

The observed vibrational frequencies are (Bolle 1982)

$$\nu_{vHCl} \approx 86.5 \times 10^{12}\,Hz\,, \quad \nu_{vCO} \approx 65.1 \times 10^{12}\,Hz$$

this corresponds to wavelengths of

$$\lambda_{vHCl} \approx 3.47\,\mu m \quad\text{and}\quad \lambda_{vCO} \approx 4.61\,\mu m$$

or to the spectroscopic wave numbers 2170 cm^{-1} and 2884 cm^{-1}.

The classical spring constants follow from Eq. (2.16) as

$$f_{HCl} \approx 478\,kg\,s^{-2} \quad\text{and}\quad f_{CO} = 1907\,kg\,s^{-2}\,.$$

This characterizes the difference in the stiffness of binding of these two molecules.

A series of vibrational energy levels is indicated by horizontal lines in Fig. 2.4 within the limits dictated by the Morse function. These modes of vibration (represented by eigenfunctions) can change only in jumps from one to the next lower or higher state. For the lowest few levels these wave functions of the vibrational modes are drawn in Fig. 2.4, the square of these functions represent the probability of finding the two atoms at an interatomic distance R between the Morse limits. Formula (2.17) indicates a linear increase of the energy with increasing quantum number v. In reality the vibration is never an exact harmonic oscillation and the vibrational energies (2.17) have to be modified by adding higher order terms which are proportional to

$$-(v+\tfrac{1}{2})^2 \quad\text{and}\quad +(v+\tfrac{1}{2})^3 \ldots$$

with strongly decreasing coefficients, the anharmonicity constants, so that the energy levels corresponding to high quantum numbers v become densely crowded below the dissociation energy D. The lowest vibrational energy corresponding to the vibrationless state ($v = 0$) is called the zero point energy

$$W_{v=0} = \tfrac{1}{2}h\nu_v - \tfrac{1}{4}xh\nu_v + \tfrac{1}{8}yh\nu_v\,,$$

with the unharmonicity constants x and y for the quadratic and cubic terms respectively. By far the greater portion of the molecules of a given gas, liquid or solid, is in the $v = 0$ vibrational level at room temperature. Inspection of the Boltzmann formula (2.2) shows that the ratios of the numbers of molecules in successive energy levels for vibrational states (energy differences typically $\Delta W_v \approx 0.1$ eV) are very small as compared to those for the lower rotational states (typically $\Delta W_r = 0.001$ eV) at room temperature ($kT \approx 0.025$ eV)

$$\left(\frac{N_{m+1}}{N_m}\right)_v \ll 1\,, \quad \left(\frac{N_{m+1}}{N_m}\right)_r \approx 1\,. \tag{2.18}$$

Therefore it is reasonable that any vibrational transition is accompanied by a rotational transition. From another point of view it is also reasonable that a rotating molecule does not exactly behave like a rigid rotator, but that it will likewise simultaneously vibrate. The quantum-mechanical selection rule for diatomic molecules and for the longitudinal vibration mode of poly-atomic molecules states that simultaneously with a transition $\Delta v = \pm 1$ there is always a transition $\Delta J = \pm 1$, except if the molecule possesses angular momentum about the axis joining the nuclei; the only stable diatomic molecule showing this type of spectrum in the near infrared is NO, and it is caused by its odd number of electrons. Hence as a rule each vibrational transition frequency is split up into a series of spectral lines with mutual separations, approximately corresponding to the respective rotational line frequencies. Figure 2.5 shows schematically the transitions between the lowest five rotational levels which accompany the vibrational transition $v = 0$ to $v = 1$ by absorption of radiation in the vicinity of the pure vibrational transition frequency $\nu_{v(0-1)}$. Two branches can be recognized, one for $\Delta J = +1$, the R-branch, and for $\Delta J = -1$, the P-branch. The line frequencies of the vibration-rotation spectrum are therefore approximately according to

$$\nu_{vr} = \nu_r \begin{cases} +\dfrac{2B}{h}(J+1) & J \to J+1 \\[2ex] -\dfrac{2B}{h}J & J \to J-1 \end{cases} \text{for} \qquad (2.19)$$

In a more accurate analysis, the increased moment of inertia, i.e. the decreased rotational constant B in the vibrationally excited state as compared to the ground state and consequently a non-linear arrangement of the lines in the R and P branch has to be considered (e.g. Wheatley 1968).

The strength of the individual lines, i.e. the absorption coefficients at the various line frequencies, are dictated by the populations of the various states. For vibrational states the first part of Eq. (2.18) and the reasoning which led us to Eq. (2.3) make it evident that only transitions between $v = 0$ and $v = 1$ are very strong.

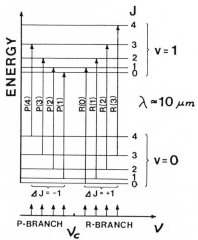

Fig. 2.5. Vibrational-rotational transitions between the lowest vibrational and rotational levels with their respective designations. (Svanberg 1978)

As far as the rotational states are concerned, the $(2J+1)$ fold degeneracy of the J-levels must be taken into consideration. $2J+1$ is the number of positions that the angular momentum vector can assume with reference to a given direction, e.g. in an external field. The relative populations N_J/N of the levels, where N is the total number of molecules per unit volume, are given by the proportionality (Fig. 2.6a)

$$\frac{N_J}{N} \propto f(J) = (2J+1)\exp[-BJ(J+1)/kT] \ . \tag{2.20}$$

Here the rotational constant B is taken constant for the different J-transitions. The shape of the vibrational-rotational absorption spectrum reflects the populations and the temperature according to Eq. (2.20). Figure 2.6b illustrates the asymmetry of the line strengths between the P- and R-branch due to the different populations which are functions of the ratio B/T of rotational constant and temperature. There are more molecules available in the lower rotational state which can make an energy jump into a higher, less occupied state than there are available for the opposite change within the rotational levels. Increasing B/T produces increasing asymmetry and can therefore be used for determining the temperature.

For the first few J-values — as long as $BJ(J+1) < kT$ — an increase of J yields an increase of the population according to Eq. (2.20) and therefore stronger lines; as soon as $BJ(J+1) > kT$ the exponential function suppresses the effect of the $(2J+1)$ factor and the line strengths decrease rapidly with increasing rotational quantum numbers J. The pure vibrational frequency ν_v does not appear at all.

Poly-atomic molecules possess much higher degrees of freedom with regard to vibrational and rotational motions. For example, the CO_2 molecule can vibrate in four different modes (Fig. 2.7a), two of which, however, will be equivalent from the point of view of energy and correspondingly will exhibit the same vibrational frequency (ν_2); this mode is twofold degenerate.

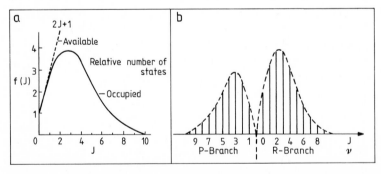

Fig. 2.6. a Relative occupation of the various rotation levels, typical example; **b** Vibrational-rotational transition band of a diatomic molecule with P- and R-branches. For a given molecule the asymmetry is a measure of the temperature. (Svanberg 1978)

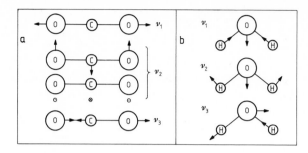

Fig. 2.7. The normal vibrational modes of **a** carbondioxide, **b** water-vapour

Because of the symmetry of the CO_2 molecule and the symmetric vibration of mode v_1, there is no electric dipole moment present by which this vibrational mode could interact with the electric field of radiation. Therefore mode v_1 of CO_2 is inactive in the infrared spectrum. In mode v_3 the distances between the atoms vary in opposite phase, therefore an electric dipole moment is created which can interact with the radiation. Because of its longitudinal oscillation, the same selection rule as for the diatomic molecule forbids the transition at exactly the vibration frequency ($\Delta v = \pm 1$, $\Delta J = 0$), see Fig. 2.8. However, mode v_2 performs an internal rotational motion while vibrating. No change of the rotational quantum number J (rotation of the molecule as a whole) is necessarily required to occur simultaneously to $\Delta v = \pm 1$ for this mode.

Therefore a large absorption peak at v_v appears as a result of the superposition of the $\Delta v = 1$ transitions at all available J-levels. This absorption region is called the Q-branch.

At this point we should make a comment on the unit cm^{-1} of the spectroscopic "wavenumber" which is used instead of frequency and which is different from the wavenumber defined in Eq. (1.15) by a factor of $2\pi 10^2$. Spectroscopists have always used wavelengths in order to characterize the colour of light, however, the proportionality between photon energy and frequency forces everybody to the use of frequency. The historical reason for the spectroscopists' definition of the wavenumber is that they normalized the frequency v (in Hertz) to the speed of light c (in $cm\ s^{-1}$), and consequently the unit of the wavenumber appears at 3×10^{10} Hz, corresponding to 1 cm wavelength. Therefore, instead of, e.g., the wavelength $\lambda = 10\ \mu m = 10^{-3}\ cm$ (3×10^{13} Hz), the corresponding frequency is given as "wavenumber" $10^3\ cm^{-1}$.

Another molecule of considerable importance in the earth's atmosphere, water vapour H_2O, is also a three-atomic molecule, however, not linearly arranged as CO_2. The axes joining the two hydrogen nuclei to the oxygen form an angle of about $104°$ (instead of the $180°$ in the case of CO_2). Therefore, the water vapour molecule has geometrically different modes of vibration and rotation, hence much more complicated rotational components ($\Delta J = \pm 1$) superimposed on the vibrational ($\Delta v = +1$) absorption. In

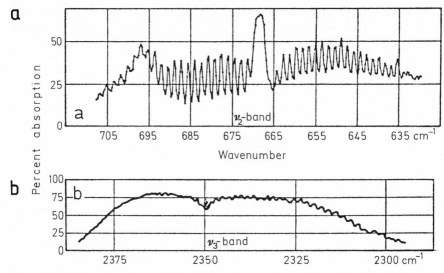

Fig. 2.8 a, b. Vibrational spectra of CO_2. **a** The ν_2 band for the perpendicular mode with a central Q-branch; **b** the ν_3 band for the longitudinal mode without a Q-branch. (Wheatley 1968)

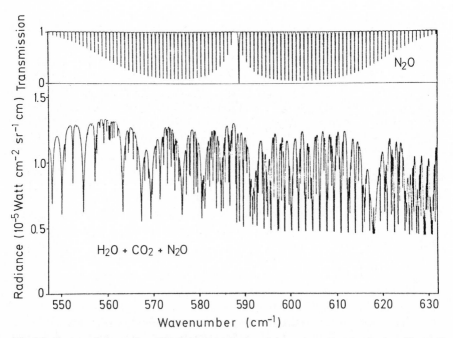

Fig. 2.9. Computed absorption of N_2O (transmission) and the computed atmospheric radiance of the superposition of H_2O, CO_2 and N_2O lines. (Bolle 1982)

Fig. 2.7b the normal modes of vibration of the water molecule are indicated. In this case there exist only three different modes without any degeneracy. For the purpose of illustration, in Fig. 2.9 a comparison is shown of the regular spectrum of the symmetric linear N_2O (in transmission) and the radiance of the superposition of N_2O with H_2O and CO_2 in a small range of wavelength at about 15 to 20 μm.

The highly irregular spectrum is mainly due to the water vapour. Water vapour has a large electric dipole moment because of its asymmetry (1.9 Debye) and a wide range of rotational levels is excited.

2.2 Widths of Spectral Lines

Up to now we have discussed the spectral lines as if they were infinitesimally narrow at exactly defined frequencies. There are, however, a number of reasons which cause line broadening and hence a certain indefiniteness of the transition frequency.

Due to the finite natural lifetime Δt_N of an excited state, (2.4), the energy of this state is determined with a finite accuracy $h\Delta v_N$ according to the Heisenberg uncertainty principle. This finite accuracy is reflected in the natural radiation width of the spectral line Δv_N which is inversely proportional to the natural lifetime Δt_N; namely

$$\Delta v_N = \frac{\Delta W}{h} \approx \frac{1}{2\pi\Delta t_N} \, . \tag{2.21}$$

The quantum mechanical explanation of this indefiniteness is the disturbance of the molecule by zero-point oscillations of electromagnetic fields which are always present in free space and able to interact with the molecular dipole moment. For typical lifetimes of electronically excited states one obtains $\Delta v_N \approx 10$ MHz. For transition frequencies below about 10^{13} Hz, (e.g. a rotational line at 3×10^{10} Hz, where the zero-point field would cause a linewidth of only 10^{-7} Hz), the thermal radiation contributes more strongly to the electromagnetic fields than the zero-point field, causing a line broadening of about 4×10^{-5} Hz. This is still quite negligible in comparison with what is caused by other types of broadening.

The finite temperature of the gas causes the molecules to move in random directions and at velocities obeying statistical laws. The probability that a molecule in a gas at temperature T has a velocity v in a particular direction is proportional to

$$\exp[-m_M v^2 / 2kT] \, , \tag{2.22}$$

where m_M is the molecular mass.

When the molecule is emitting a photon while moving with velocity v, an observer in that particular direction receives a radiation which is shifted in frequency by an amount of

$$\delta v = \pm \frac{v}{c} v_0 \; , \tag{2.23}$$

the Doppler shift, where v_0 is the resonant frequency without Doppler shift. The $\pm$ signs apply for the observer in the forward or backward direction respectively. Vice versa the Doppler effect causes the moving molecule to absorb out of the radiation a frequency which is shifted by the same amount (2.23) with respect to v_0. The spectral line can be represented by the absorption coefficient κ which is a measure of the ability of a given medium to absorb a certain portion of the radiation intensity per unit path. In the vicinity of a spectral line κ is strongly dependent on frequency. Because of the randomly oriented motions of the large number of absorbing molecules in a volume of gas the resulting absorption of the radiation intensity becomes a bell-shaped function about the exact transition frequency v_0. The absorption coefficient of the Doppler-broadened line can easily be found by inserting Eq. (2.23) into (2.22)

$$\kappa_D(v) = \kappa_{D0} \exp \left[-\frac{m_M c^2}{2kT} \left(\frac{v - v_0}{v_0} \right)^2 \right] \; , \tag{2.24}$$

where κ_{D0} is the absorption coefficient at line centre ($v = v_0$), as evident from Eq. (2.24).

The spectral line due to Doppler broadening has a Gaussian shape centered at v_0. The line is consequently symmetric and at half the maximum value it has a half-width [see also Eq. (3.54) and Fig. 3.18] of

$$\Delta v_D = \frac{v_0}{c} \sqrt{\frac{2kT}{m_M} \ln 2} = v_0 \xi \sqrt{\frac{T}{M}} \; , \tag{2.25}$$

where M is the molar weight in grams, i.e. the weight of one molecule in grams times Avogadro's number (6.023×10^{23}), or in simpler terms: M is the relative molecular weight in grams, e.g. for H_2O approximately 18 g. In $\xi = 3.58 \times 10^{-7}$ degree$^{-1/2}$ all constants are gathered and T is the absolute temperature in Kelvin. At $T = 300$ K for simple molecules, e.g. CO_2, HCl, Eq. (2.25) is approximately $\Delta v_D \approx 10^{-6} v_0$. Higher temperatures cause enforced motions of the molecules, hence more broadening of the absorption. Heavier molecules are more inert against thermal agitation and consequently show less broadening.

Figure 2.10 shows the range of line widths due to this approximation over a wide range of transition frequencies. The effect of temperature in the atmosphere on the line width corresponds to the $T^{1/2}$-dependence, therefore accurate measurements of the linewidth allow deduction of the temperature of the particular regions viewed by a remote spectrometer. Evidently neither the total atmospheric pressure nor the density of the particular species of which a spectral line is measured enter the formula of the Doppler linewidth.

This is different in the third line-broadening mechanism to be discussed here, collisional- (or pressure-)broadening. At atmospheric pressures down to less then 100 millibar, i.e. up to at least stratospheric heights, the broadening

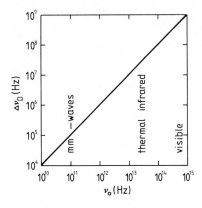

Fig. 2.10. Typical width of Doppler broadened lines of light molecules at room temperature as a function of the respective centre frequency

of all vibrational lines of the atmospheric constituents is dominated by collisional effects. For most pure rotational lines this limit is at less than 1 mbar, i.e. at mesospheric heights. The oscillations of the molecules, i.e. the particular vibrational and rotational states, are interrupted by collisions with neighbouring molecules due to their density and their thermal agitations. This means that the life-times of all dynamical states of the molecules are very much shortened and therefore the linewidth very much broadened. The theory assumes that after a collision, when the oscillation has stopped, it starts again with a phase having no relationship to the phase before the collision, i.e. strong collisions are assumed. When applied to the case of rotating molecules, this means that the orientation of the molecules after collision is random. Due to the collisions within the gas volume with given physical conditions, i.e. temperature and pressure, an equilibrium will be reached between the thermal energy of the molecular motions and the electrical energy of the dipole oscillations. This means that the population of the various energy levels of the molecules very closely obey the Boltzmann energy distribution, hence the spectral properties of the absorption lines reflect the kinetic temperature of that gas volume.

Derived from the theory indicated above, the lineshape (the absorption coefficient as a function of frequency) according to van Vleck-Weisskopf (Townes and Schawlow 1955) can be found as

$$\kappa_c(v) = \kappa_{c_0} \frac{p}{T} v^2 \frac{\Delta v_c}{(v - v_0)^2 + (\Delta v_c)^2} \ . \tag{2.26}$$

The half-width due to molecular collisions Δv_c is a function of pressure and temperature and can be represented by

$$\Delta v_c = \Delta v_{c_0} \frac{p_e}{p_{e0}} \left(\frac{T_0}{T} \right)^n , \tag{2.27}$$

where Δv_{c_0} is the linewidth at normal pressure p_{e0} and normal temperature T_0. The value of Δv_{c_0} varies for different molecules between 0.3 and 3 GHz. The lower value applies approximately for high rotational transitions of water vapour, the higher value for ozone. In a mixture of various component gases the pressure effective for the line broadening is given by $p_e = p + (B-1) p_i$, where p is the total pressure, p_i the partial pressure of the absorbing gas i and B is the self-broadening coefficient of that particular gas. The exponent n in Eq. (2.27) is related to the dependence of the effective collisional cross-section on the velocity of the collision partners, i.e. on the temperature. The value of n varies between 0 and 1, e.g. for the transitions between the lowest rotational levels of water vapour n varies between ≈ 0.3 and ≈ 0.65 (Waters 1976), with a mean value of approximately 0.5.

For high pressure ($\gtrsim 100$ mbar, i.e. in the troposphere) where the line-width Δv_c of rotational lines can become comparable to the line centre frequency v_0 itself, the line shape (2.26) becomes asymmetric (see also Gordy and Cook 1970, p. 49).

Figure 2.11 shows a laboratory measurement confirming the asymmetric van Vleck-Weisskopf line shape over a wide frequency range of a microwave rotational line in the high pressure regime with a high degree of water vapour content. From this example we see that the pressure-broadened linewidth can amount to a considerable fraction of the line centre frequency itself, if measuring in tropospheric heights.

Because of the approximately exponential decrease of pressure with increasing height we shall meet a region where the pressure broadening is smaller than the Doppler broadening effect which is independent of the pressure. In this transition region the resulting linewidth is approximately

$$\Delta v \approx [(\Delta v_D)^2 + (\Delta v_c)^2]^{1/2} \ . \tag{2.28}$$

The ratio of the linewidth due to Doppler and due to collisional broadening is, in a rough approximation for average molecules (with a relative mass of the order of 50) and at a temperature $T \approx 300$ K, given by

$$\frac{\Delta v_D}{\Delta v_c} \approx 10^{-12} \frac{v_0}{p} \ , \tag{2.29}$$

where v_0 is the centre frequency of the line in Hertz and p is the atmospheric pressure in millibar. This ratio increases for lighter molecules as $M^{-1/2}$ (molecular mass M), and it depends on temperature as T^x where x assumes typically values between 0.5 and 1.

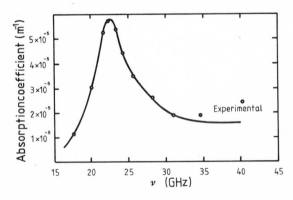

Fig. 2.11. Absorption coefficient of water vapour in air (10 g H_2O m^{-3}). Comparison between measurement and the van Vleck-Weisskopf line shape. (After Becker and Autler 1946)

Formulae (2.27) and (2.28) are very approximate indeed; a general formula for a line shape with pressure and Doppler broadening has been derived by Voigt (see, e.g. Penner 1959) and has to be applied for more exacting quantitative interpretations.

Figure 2.12 shows schematically the linewidth as a function of the height in the atmosphere for a fine structure line of the oxygene molecule at millimetre waves (about 118 GHz) and a vibrational line of carbondioxide in the "thermal" infrared, approximately $670 \, \text{cm}^{-1}$. Due to the fact that the

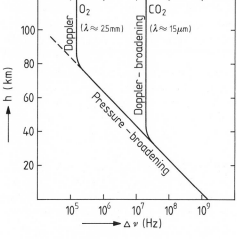

Fig. 2.12. Approximate relationship between atmospheric height h and linewidth for a microwave line of O_2 and an infrared line of CO_2 (idealized isothermal atmosphere and equal $\Delta \nu_{c0}$-values for O_2 and CO_2 are assumed)

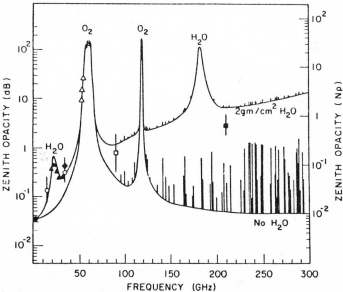

Fig. 2.13. Atmospheric zenith opacity in the microwave spectral region due to O_2 and O_3, the effect of H_2O-vapour is shown by the two different curves. (Waters 1976)

Doppler broadening is proportional to the line centre frequency, this portion is more than two orders of magnitude smaller for the O_2 line as compared to the CO_2-line and as a consequence the transition from pressure to Doppler regime is approximately 40 km higher up in the atmosphere. If one is able to measure remotely the linewidths for different layers of the atmosphere separately, this can yield information on the respective layer from which the line radiation is originating. We shall see in Chapter 5 that this is an important tool in strato- and mesospheric sounding.

Figure 2.13 illustrates the contributions of the different heights, i.e. different pressures to the linewidth for the microwave zenith oriented observation. The O_2- and H_2O-contributions are dominated by tropospheric pressures, while the O_3-lines stem from the stratosphere and, on this scale, are unresolvably narrow. The concept of opacity will be discussed in some detail and utilized in the last chapter. In the present context we may regard it approximately as the product of absorption coefficient and effective path length, e.g. thickness of the atmosphere. Thus the opacity represents the absorption coefficient in at least a qualitative way.

2.3 Applications to the Earth's Atmosphere

In remote sensing of the earth's surface, the atmosphere is a disturbing medium, which impedes the correct measurement of the surface features over a wide part of the electromagnetic spectrum. In some spectral regions, e.g. the far infrared around 100 micron wavelength, the presence of many intense absorption lines caused by the atmospheric constituents even prevents any useful remote sensing. On the other hand, the spectral lines characterize the absorbing species in an unambiguous way and the intensities and widths of the lines reflect the physical parameters (temperature, density, total number) in the observed volume. Thus the spectral lines are the important key to remote sensing of the atmosphere itself.

One of the most important constituents of the atmosphere is water vapour − not only for its determining effect on weather development within the troposphere, but also as an important partner in the photochemical reactions in strato- and mesosphere and as an important agent of the heat exchange and of the atmospheric motions. As indicated in connection with Fig. 2.9, water vapour exhibits a very large number of absorption lines apparently in complex distribution over the infrared and microwave spectrum. The reason for this is the bent structure of this molecule.

Figure 2.14 shows the molecular dimensions and indicates the rotational modes. The structure of water vapour is classified as an asymmetric top with the three rotational constants 835.7 GHz, 434.9 GHz and 278.4 GHz. The vibrational modes, presented in Fig. 2.7 b, are accompanied by the rotations of the molecule and accordingly vibrational-rotational absorption spectra are found (see Table 2.3).

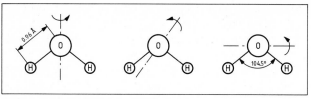

Fig. 2.14. Dimensions and rotational modes of the water vapour molecule, its electric dipole moment is approximately 1.9 Debye. (Townes and Schawlow 1955)

Table 2.3. Vibrational-rotational spectra of the normal and mixed vibrational modes of water vapour. (After McClatchey et al. 1973)

Central wave-number [cm^{-1}]	Spectral range [μm]	Transition upper''-lower' vibrational level $(v_1\, v_2\, v_3)'' - (v_1\, v_2\, v_3)'$	Normal modes (see Fig. 2.7b)
1595	4.2 − 7.8	010 − 000	v_2
3657	2.5 − 2.8	100 − 000	v_1
3756	2.2 − 3.3	001 − 000	v_3
5331	1.6 − 2.2	011 − 000	−
7250	1.25 − 1.6	101 − 000	−
8807	0.9 − 1.25	111 − 000	−

Water vapour also exhibits a pure rotational spectrum of significant importance while the molecule remains in the vibrational ground state. It is centered at about 50 μm and it extends from about 10 μm to a little more than 1 cm wavelength. For the purpose of illustration in Fig. 2.15a the atmospheric transmittance in the wavelength range 26 μm to 31 μm (wavenumbers 380 to 320 cm^{-1}) is shown for a horizontal transmission path of 10 km length at 12 km altitude. At sea level the widths of the lines are much larger due to the higher pressure, therefore only a few narrow regions remain with reasonable transmittance.

Another important constituent of the atmosphere causing strong absorption in the far infrared region is (as already indicated) carbon-dioxide. We discussed in Section 2.1 the vibrational modes of the CO_2-molecule, which correspond to the following vibrational centre frequencies of the normal modes (in units of wavenumbers)

$$v_1 \approx 1340\;\mathrm{cm}^{-1}, \qquad v_2 \approx 667\;\mathrm{cm}^{-1}, \qquad v_3 \approx 2349\;\mathrm{cm}^{-1}$$

(Wheatley 1968). Mode v_1 is inactive in the IR-absorption spectrum due to the missing dipole moment. Vibrational-rotational spectra centred at the v_2 and the v_3 mode of vibration are shown in Fig. 2.8.

In contrast to water vapour, the CO_2 molecule already exhibits at longer wavelengths a vibrational-rotational absorption band centred at 667 cm^{-1} ($\lambda \approx 15$ μm). Figure 2.15b shows a small part of the short wavelength side of this absorption spectrum for the same observing geometry as for H_2O in

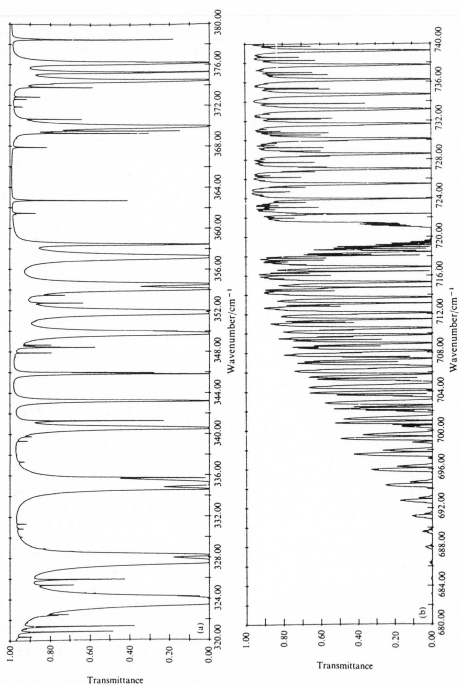

Fig. 2.15. Atmospheric transmission for a 10 km horizontal path at 12 km altitude as computed by McClatchey and Selby (1972) for a the 31 to

Fig. 2.15a. Up to wave numbers 700 cm^{-1} the atmosphere is opaque, the lines in this spectral region are mainly due to CO_2.

Excepting those two atmospheric constituents discussed so far, there are many more which contribute to the attenuation of the radiation due to line absorption in the important wavelength ranges of the near infrared ($\lambda \approx 1$ μm) to the thermal infrared ($\lambda \approx 10$ μm). Figure 2.16 summarizes the absorption spectra of the most important agents, showing the absorption along a vertical path through the whole atmosphere. The individual effects are added in the lowest part of the diagram. It makes obvious that only one broad region between 8 and 12 μm wavelength remains transparent, while below 4 μm there

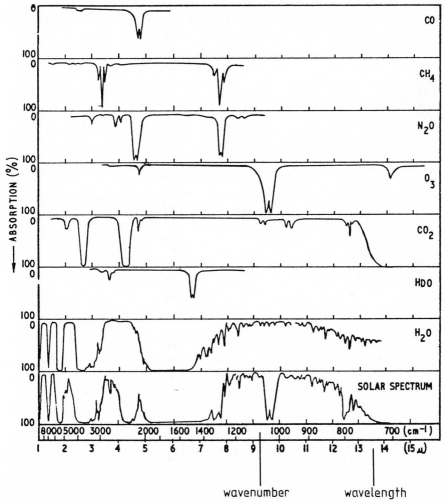

Fig. 2.16. Infrared absorption spectra of various atmospheric constituents separately and their combined effect. (Valley 1965)

are four narrow bands which can be utilized for remote sensing through the atmosphere. This summary diagram also confirms that the atmosphere is opaque in the far infrared beyond about 14 μm. Only at much longer wavelengths (at about 1 mm) the atmosphere again becomes more and more transparent, because only fewer and weaker rotational transitions cause absorptions.

The ozone molecule which is abundant mainly in stratospheric heights contributes considerably to the total absorption in only one small infrared wavelength region. Ozone, however, absorbs much more in the ultraviolet and – as we have already seen in Fig. 2.13 – also in the millimetre wave range. A few more remarks on these features of ozone will be made below.

Figure 2.17 is another summary diagram. It indicates the absorption effects of the relevant atmospheric molecules on the solar spectrum as received at the earths surface. The computed radiance (Planck's function) of a blackbody at 5900 K, a temperature which fits the solar irradiation curve outside the atmosphere, is compared with the irradiance at the earth's surface. The individual absorption (as far as distinguishable) is marked.

The most abundant molecules in the air, N_2 and O_2, have no electric dipole moment because of their symmetric charge distribution. They therefore show no vibrational nor simple rotational spectra as discussed so far. However, the oxygen molecule contains two of the orbital electrons which are unpaired, resulting in a permanent magnetic dipole moment. The molecule rotates end over end with quantized angular momentum. The rotational quantum number N describing the state of quantization may take odd values only

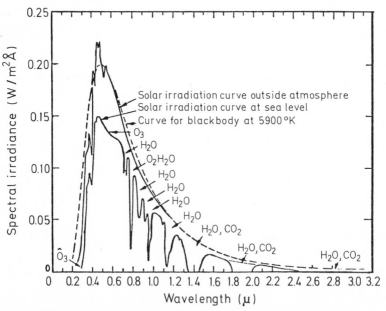

Fig. 2.17. The solar spectrum between the ultraviolet and the infrared. The molecules causing the respective absorption features are marked. (Valley 1965)

($N = 1, 3, 5 \ldots$). This is because only the corresponding anti-symmetric rotational wave functions can compensate for the anti-symmetric wave function due to the unpaired electrons, in order to satisfy the overall symmetry requirements. The spin angular momentum of the unpaired electrons is described by the quantum number S. The value is $S = 1$ because two spins, each representing $S = \frac{1}{2}$, are added up. The two momenta combine to form a total angular momentum vector J, which is the vectorial sum of N and S. Consequently, for every N state there are three J states depending on the relative orientation of S with respect to N.

$$J = \begin{cases} N-1 & \text{for} \quad S = -1 \\ N & \text{for} \quad S = 0 \ . \\ N+1 & \text{for} \quad S = +1 \end{cases} \tag{2.30}$$

This means that the rotational levels of the O_2 molecule are each split into a spin triplet corresponding to the three possible space orientations of the electronic spin vector S. The energies of these levels differ only slightly resulting in a so-called fine structure of the rotational levels. For most of the other atmospheric molecules all spins are paired, ($S = 0$), the actions of the spins on the molecular transitions are cancelled, hence $N = J$ all time, a case which is called singlet state.

In a triplet state, a molecule without electric dipole like O_2 can make magnetic dipole transitions between these split rotational levels according to the selection rules (Gordy and Cook 1970) along two series

1. simultaneously: $\Delta N = 0$ and $\Delta J = \pm 1$

2. simultaneously $\Delta N = \pm 2$ and $\Delta J = 0, \pm 1$. $\qquad$ (2.31)

Series 1 represents transitions between triplet components of given rotational levels. For O_2 this series falls in the wavelength region near 5 mm (60 GHz) all very close together, because all the $\Delta J = \pm 1$ transitions for the different N-values correspond to very close energy differences. There is one exception: for $N = 1$ the transition $J = 0 \leftrightarrow 1$ appears at about 2.5 mm wavelength (118 GHz) as an isolated line. The transitions $(J = N) \leftrightarrow (J = N \pm 1)$ are tabulated by Waters (1976) up to $N = 39$ (only odd N-values) and except the one mentioned, all are concentrated between about 49 GHz and 70 GHz.

Magnetic dipole transitions as described here for O_2 are typically 10^{-4} less intense than electric dipole transitions, but because of the large abundance of oxygen in the atmosphere, O_2 produces quite strong atmospheric absorptions by these fine structure lines (see Fig. 2.13). At tropospheric pressure the collisional broadening causes an overlap of these lines. At sea level the linewidth is about 1 GHz, while the separation between neighbouring lines is typically half that amount. The effect of pressure on the width of these lines is demonstrated by the computed attenuation spectrum in Fig. 2.18.

The rotational constant for O_2 is $B = 43.1$ GHz and the normal vibrational frequency is at $\nu_1 \approx 1580 \text{ cm}^{-1}$ (Townes and Schawlow 1955) about an equilibrium bond length of 0.1207 nm (Gordy and Cook 1970).

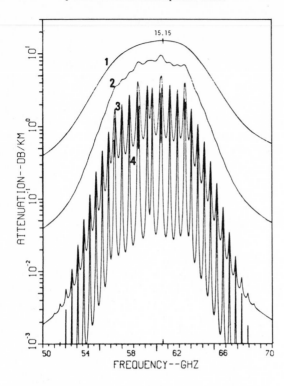

Fig. 2.18. Attenuation (in decibels per kilometre) for the oxygen microwave band. The curves are computed by Liebe (1981) for different altitudes h and relative humidities m.
1 $h = 0$ km, $m = 50\%$;
2 $h = 10$ km, $m = 5\%$;
3 $h = 20$ km, $m = 1\%$;
4 $h = 30$ km, $m = 0$

The second series of Eq. (2.31), with $\Delta J = 0$, represents pure rotational transitions, the first of these lines (for $N = 1 \to 3$ and $J = 2$) appears at 424.7 GHz. The second series, with $\Delta J = \pm 1$, represents combined transitions between molecular-rotational and electron-spin states. For O_2 this series originates also in the submillimetre wave region (at 367 GHz) and extends into the infrared.

Due to the fact that the transitions of O_2 are based on its magnetic dipole moment, there appears an added complication, in particular when measured at the higher levels of the atmosphere (mesosphere) where the linewidth is very narrow. In the presence of a magnetic field (the earth's magnetic field of approximately 0.5 Gauss is already sufficient for this) each J-level is split up into M levels due to the Zeeman effect, as illustrated for the simplest case in Fig. 2.19.

The resulting components are polarized and their number depends on the rotational states involved and on orientation with respect to the earth's magnetic field. The anisotropic interaction between the magnetized molecule and radiation requires circularly polarized waves for coinciding directions of observation and magnetic field. The appearance of positive or negative circular polarization depends on the sign of the Zeeman effect, for which $\Delta M = \pm 1$ adds to the previously discussed fine structure transition. For an observing direction perpendicular to the magnetic field, linear polarized waves

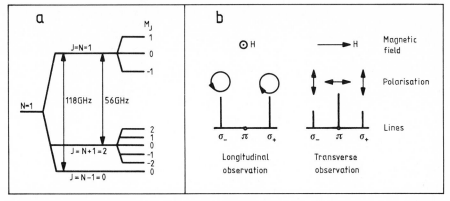

Fig. 2.19. Zeeman splitting of O_2 energy levels (Croom 1978) **a** the level scheme for $N = 1$; **b** the lines and the polarizations of the corresponding radiation

can cause an interaction. Figure 2.19b indicates the spectral lines for both situations, the designations being σ_+, σ_- and π for $\Delta M = +1$, -1, 0, respectively. The π-component does not appear at all in the longitudinal and it shows up as the strongest line in the transverse configuration. The frequency shift of the σ-components with respect to the unshifted line (the π-component) is

$$\delta \omega_z = \pm \frac{1}{2} \frac{e}{m} B \, , \tag{2.32}$$

with the electron charge e and mass m and the magnetic induction B in $Vs\,m^{-2}$; in more practical units we can write $\delta v_z(MHz) = \pm 1.4 \, B$ (Gauss). Since the earth's magnetic field is about 0.5 Gauss, for the atmosphere the resulting line frequencies become approximately $v = v_0 \pm 0.7$ MHz. This means that extra line broadening due to Zeeman effect can become a disturbing factor for mm-wave lines above about 50 km (cf. Fig. 2.12), while separation of the Zeeman splitted lines becomes noticeable above about 60 km altitude and will depend on the exact strength and orientation of the earth's magnetic field.

The methane molecule, CH_4, is another example of a molecule lacking any electric dipole moment. In this case the reason is the perfect (tetrahedral) symmetry in the vibrational ground state. Therefore, CH_4 is classified as a spherical top molecule, and there are no pure rotational spectral lines. Because of its symmetry, CH_4 has only one non-degenerate, one doubly-degenerate and two triply-degenerate vibrations (Herzberg 1945). One of the triply-degenerate states (v_3) is the one where all four H-atoms oscillate synchronously in one direction with respect to the C-atom, while the whole molecule rotates (Fig. 2.20a).

This combined motion causes Coriolis forces which produce an interaction between mutually degenerate vibrations (e.g. between x- and y-directions

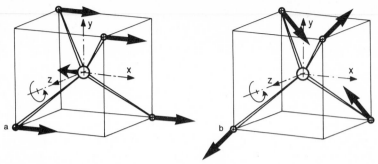

Fig. 2.20a, b. Two triply-degenerate modes of vibration of the tetrahedral CH_4-molecule. **a** The v_3-vibration assumed parallel to the x-direction; **b** the v_4-vibration. The rotation is assumed around the z-axis in both cases

when z is the axis of rotation). As a consequence, the degeneracy is split and the result is a fine structure of lines about the fundamental band v_3 at 3020 cm^{-1} ($\lambda = 3.31$ μm). The spacing of the lines due to transitions between the rotational levels in the P- and R-branch is approximately 10 cm^{-1}. The second triply-degenerate vibration (v_4) with anti-symmetric oscillations of the H-atoms (Fig. 2.20b) is at 1306 cm^{-1} ($\lambda = 7.65$ μm), the spacing of the lines being approximately 6 cm^{-1}. This spectrum is less regular because of strong Coriolis interaction with the (IR-inactive) doubly-degenerate vibration at a nearby frequency. Molecular parameters determined from these spectra are rotational constant $B \approx 5.25$ cm^{-1} (157.5 GHz) and from this follows the moment of intertia $I \approx 0.53 \times 10^{-46}$ kg m^{-2} and the C$-$H distance 0.109 nm. The absorption features due to these two vibration-rotation bands of atmospheric CH_4 are indicated in Fig. 2.16.

Another method to remove the symmetry of CH_4 and to induce an electric dipole moment is applicable if the molecule can become polarized by an intense coherent radiation (Laser). This technique is called Raman spectroscopy and will be discussed in Chapter 4. By this method also the non-degenerate v_1 (totally symmetric stretching) vibration at 2914 cm^{-1} and the harmonic of doubly-degenerate v_2 vibration ($2 v_2$ at 3071 cm^{-1}) can interact with radiation additionally to the v_3 and v_4 vibrations (Herzberg 1945).

The remote detection of CH_4 by its infrared absorptions at 1.33 μm and 1.66 μm due to $v_2 + 2 v_3$ and $2 v_3$ vibrations has recently been reported by Chan et al. (1985).

The ozone molecule is, like water vapour, an asymmetric top molecule (bond angle 116.5°, bondlength between neighbouring O-atoms 0.128 nm) with an electric dipole moment of about 0.53 Debye. Due to its asymmetry, it shows a wide and complicated spectral distribution of absorption lines. We can distinguish three spectral regions which are important in our context.

Ozone very efficiently absorbs ultraviolet solar radiation in the so-called Hartley band, a broad continuous absorption spectrum from 200 to 300 nm (Houghton 1977). Ozone is formed in the stratosphere and mesosphere (10 to

80 km) by photochemical processes, the peak of its concentration occurring at about 25 km altitude. This is the same region where most of the ultraviolet absorption by ozone takes place. At levels below 70 km the absorbed radiative energy goes into kinetic energy of the molecules, i.e. into increasing the atmospheric temperature, resulting in the stratospheric temperature peak.

A second spectral region is dominated by vibration-rotation bands in the infrared region between 3 μm and 15 μm with a particularly strong absorption feature in the 8.3 μm to 10 μm range about the centre wavenumber 1042 cm^{-1}, due to the transiton between the vibrational ground state and the v_3 vibrational mode (Bolle 1982). This absorption feature of O_3 can be recognized in Fig. 2.16.

Finally, in the millimetre and submillimetre frequency range, ozone shows a vast number of spectral lines due to electric dipole transitions between rotational states of the molecule. The stronger lines of the millimetre range (below a frequency of 300 GHz) have been tabulated by Waters (1976). A few of the most prominent ones, which are important for the remote investigation of the ozone layer, are listed in Table 2.4. The absorption coefficient κ_0 at line centre are computed ones for a temperature of 220 K, the approximate temperature of the ozone layer in the atmosphere.

Because of the increasing technical difficulty to detect weak signals at increasing frequencies, only lines with strongly increasing absorption coefficients and established confirmation in laboratory measurements have been selected for Table 2.4. Depending on the observing situation (ground-based or space-borne, spectral separation of different resonances within a sensor bandwidth, effect of temperature on the resonance), other line frequencies can be more attractive for the observation of the ozone layer (Waters et al. 1984).

In conclusion let us take a somewhat abridged view of the main absorption features of the standard atmosphere, with an assumed water vapour content, over the whole wavelength region between the visible and microwaves (Fig. 2.21).

Table 2.4. A few of the most prominent millimetre wave lines of ozone which are important for the remote detection. (After Waters 1976)

Frequencies v_0 (GHz)	Absorption coefficients κ_0 (m^{-1})
101.74	0.18
142.17	0.54
165.78	0.68
184.38	0.64
195.43	0.87
243.45	1.29
248.18	1.31
288.96	1.62

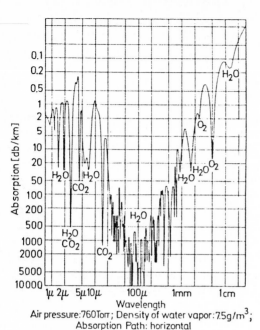

Fig. 2.21. Transmission loss in air under clear weather and standard atmospheric conditions in the infrared and millimetre spectral ranges. (Personal communication by F. Kneubuehl, ETH-Zürich 1974)

The species causing the main part of absorption are indicated without detailed association to individual line features and without a high degree of accuracy. The absorption along a horizontal transmission path is given in decibel per kilometre. We can recognize that in the far infrared region (between 20 μm and 1 mm wavelength) the atmosphere is almost completely opaque and thus prevents remote sensing of the earth's surface. The reason for this is the complex coverage of this wavelength range by the lines of water vapour. Of course the assumption of 7.5 g m^{-3} H$_2$O is arbitrary, but even for a considerably lower water vapour content the absorption of radiation in this specific wavelength region would be too high for transmission over any useful distance. Another feature is the high transparency of the atmosphere for microwaves, i.e. for wavelengths longer than about 1 cm, as compared to the infrared and optical wavelengths. The largest wavelength in the infrared useful for observations through the atmosphere is obviously around 10 microns, the "thermal" infrared band.

A final remark on the units of decibel for readers not familiar with this term. The attenuation of the intensity of radiation along a given path can be presented in decibel by computing the logarithm of the ratio of the respective intensities before (I_0) and after (I) a path of a given length in a given direction.

$$\text{Attenuation (dB)} = 10 \ _{10}\log \left(\frac{I_0}{I} \right) . \tag{2.33}$$

Table 2.5. Approximate attenuations in decibel for a wide range of intensity ratios

I/I_0	1	0.8	0.5	0.1	0.01	0.001
Attenuation (dB)	0	1	3	10	20	30

Table 2.5 relates a few intensity ratios I/I_0 and the corresponding attenuation values in decibel.

Referring to Figs. 2.18 and 2.21, where the attenuation due to absorption over 1 km horizontal path is given, we recognize that for a total attenuation less than 1 dB the atmosphere is almost transparent, while for attenuations more than 20 dB it may be regarded as opaque.

The solar radiation has its spectral peak in the visible region and can enter after only a weak attenuation (see Fig. 2.17). It heats the soil and the lower layers of the atmosphere up to approximately 300 K. But the thermal radiation of the earth at this temperature is concentrated between 5 μm and 100 μm (see Fig. 2.2), a wavelength range which partly ($\lambda \gtrsim 14$ μm) prevents the radiation energy from leaving the atmosphere because of absorption.

3 Spectral Properties of Condensed Matter

3.1 Elementary Theory of Organic Dyes

In the visible range of the spectrum we are familiar with the perception of colour, which arises if the surface of a material reflects the natural light in one part of this spectrum more strongly than in others. The light which is not reflected penetrates into the material and will eventually be absorbed. Among the most important media observed in remote sensing applications on the earth's surface are those which are composed, for a considerable part, of organic molecules.

In the previous chapter we have discussed the interaction of radiation with various modes of movement of isolated molecules. This was sensible and useful for gases, where the distance between the molecules is sufficient to allow them virtually all kinds of individual motion, but in the case of condensed matter (liquids and solids) the molecules are packed together so closely that any noteworthy proper motion is prevented by collisions with neighbouring molecules. In more physical terms, the mean free oscillation time is of the same order of magnitude or even much shorter than the oscillation period. Thus the energy levels of the vibrational and rotational states, for example, are so ill-determined that an entire range of photon energies is absorbed and only a broad absorption spectrum is obtained. We shall discuss these features in the last section of this chapter in connection with the wave dispersion by damped resonance and relaxation of the polarization. There remains to be discussed only one other dynamic mode of interaction of radiation with molecules, that of the electron cloud around and between the atoms constituting the molecule.

The absorption of ultraviolet and visible light corresponds to a disturbance of the valence electrons in the electron cloud of a molecule, resulting in the formation of an electronically excited state. According to quantum theory the possible energies of the discrete energy states W_m of the molecule are related to the wave functions ψ_m describing each of these states by the time-independent Schroedinger equation of wave mechanics

$$H\psi_m = W_m\psi_m . \tag{3.1}$$

The Hamiltonian or total energy operator H is a function of the momenta and coordinates of the particles in the molecule. The operator H operates on ψ_m, much like the position derivative d/dx and the time derivative d/dt operate on functions containing position x and time t, and does not simply multiply it by a constant factor. In electronic excitation processes we will ignore the comparatively much slower vibrational motion of the atoms in the molecule and the even slower rotational motion of the molecule itself, and take ψ_m as a function of the coordinates of the electrons only. The product of each amplitude function ψ_m with its complex conjugate ψ_m^* represents a probability function for the state m, such that the integration over all space coordinates will be unity

$$\int \psi_m \psi_m^* d\xi = 1 \tag{3.2}$$

where $d\xi$ is an infinitesimal element of volume. Wave functions that satisfy Eqs. (3.1) and (3.2) are said to be normalized eigenfunctions, and their corresponding (discrete) energies W_m are called eigenvalues. The relation between the eigenfunctions and the probabilities of finding a particle in a certain state has already been mentioned when the vibrational modes of a molecule in Fig. 2.4 were discussed. If appropriate wavefunctions ψ_m can be found, solution of Eq. (3.1) allows all of the associated W_m values to be calculated.

In general, coloured molecules have particularly large molecular frameworks, and exact solutions of Eq. (3.1) are not possible for complex molecules. Thus a series of approximations has to be made before even the simplest molecules can be handled. In the context of this introductory course, only rather crude techniques will be used for the approximations.

The simplest approximation assumes that the complete electronic wave function for the ground state of a molecule can be factored out into products of less complicated wave functions ψ_0, each of which describes the behaviour of one electron only. These one-electron wave functions are the so-called molecular orbitals, important for the understanding of bonding in molecules. Each molecular orbital is considered as being generated by the superposition of the overlapping electronic orbitals ϕ_i of the participating atoms. Depending on the energy eigenvalues of the atom, there are several eigenfunctions ϕ_i', ϕ_i'', ... which correspond to different orbitals characterized by an orbital quantum number l, but because they assume the same energy they are called degenerate. The simplest orbitals are

$l = 0$ s orbital (spherically shaped distribution function)
$l = 1$ p orbital (dumb-bell-shaped distributions in three spatial directions p_x, p_y, p_z)
$l = 2$ d orbital (rosette-shaped).

The atomic orbitals, i.e. the wave functions ϕ_i, have the property of phase (+ or − sign in Fig. 3.1) and describe the spatial distribution of the probability of finding the electrons around the nuclei. From these atomic orbitals ϕ_i wave functions (molecular orbitals) can be generated, each of which has characteristic spatial properties and, as in atoms, each molecular orbital can, according to the Pauli exclusion principle, accommodate a maximum of two electrons of opposite spin. The overlapping of wave functions to form molecular orbitals is shown schematically in Fig. 3.1.

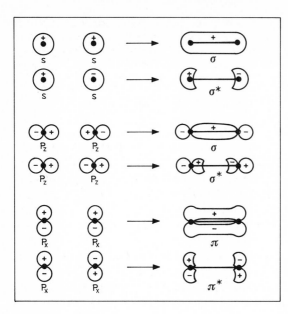

Fig. 3.1. Synthesis of bonding and anti-bonding molecular orbitals ψ_0 by in-phase and out-of-phase superposition of atomic orbitals ϕ_i. (Griffiths 1976)

In the molecule an in-phase overlap ($+ +$ or $- -$) leads to reinforcement of the composite wave function, whereas out-of-phase overlap ($+ -$) leads to cancellation. The energies of the resultant molecular orbitals are related to the degree of overlap of the component atomic orbitals. In-phase overlap of two atomic s orbitals or end-on overlap of two atomic p orbitals is particularly effective and leads to so-called σ-bonding molecular orbitals, which are cylindrically symmetric about the internuclear axis (the z-axis in Fig. 3.1). When the probability distributions are large in the internuclear region, the atoms are strongly bonded (have low energies) and represent very stable molecules only rarely taking part in chemical reactions. Out-of-phase overlap results in high energy antibonding σ^* orbitals, with very low probabilities for finding electrons in the internuclear region.

The dumb-bell shape of p_x and p_y orbitals allow a different type of overlap, as shown by the two lowest configurations in Fig. 3.1 for bonding and anti-bonding orbitals respectively. In-phase overlap of this type leads to bonding π orbitals, but this overlap is much less effective than that of σ orbitals; π orbitals are not as low in energy and do not produce strong chemical bonding. The wave functions of π orbitals have maximum values on two opposite sides of the internuclear axis and occupy a larger volume. The electrons are not as firmly held by the nuclear framework and hence are readily displaced (delocalized). Molecules containing π bonds are therefore chemically reactive.

The energy levels of bonding and anti-bonding orbitals can be illustrated by the simple ethylene-molecule with a single C = C double bond (Fig. 3.2). The level marked α is the sum of the energies of a p orbital electron in each of

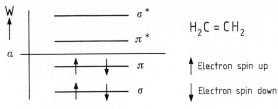

Fig. 3.2. The simple ethylene molecule and the energy level scheme of the molecular orbitals of its $C = C$ bond. The energy of the molecule before bonding takes place is arbitrarily set zero

the C atoms before bonding. The two C atoms cannot form a stable molecular bond unless the energy of the molecular orbital is lower than the energy of the original atomic orbitals. In the molecular ground state only the levels of bonding orbitals are occupied by electrons. The difference between α and π, for example, is the bonding energy of the π orbital, while the difference between α and π^* corresponds to the antibonding energy of the π^* orbital. Thus the following order of orbital energies generally holds in an organic molecule: $W_\sigma < W_\pi < W_\pi^* < W_\sigma^*$.

The large energy separation between the σ and π orbitals in general prevents mutual interaction, therefore they can be treated separately.

The π bonds play an important role for the chemical and physical properties in unsaturated compounds. Those which contain double bonds in conjugation, such as butadiene and hexatriene, are of particular interest. In conjugated olefins the π electrons are not localized between two centres, but spread out over the whole π system. As a consequence such compounds become more stable (the stability is given by the delocalization energy) and the difference in bond lengths between the double and single bonds becomes smaller. Also, the interaction of conjugated π systems with light is shifted to longer wavelengths, i.e. to lower photon energies (Meier 1963).

As a simple typical example we will discuss the 1,3-butadiene molecule, which has four π bonds in addition to nine σ bonds. Figure 3.3 shows the energy level scheme (orbital diagram) of the π orbitals in ground (a) and excited states (b, c, d), where each arrow indicates an electron and its associated spin direction. In the ground state each of lowest levels is occupied by a pair of electrons of opposite spin. In the electronic configurations b and c the excited electron retains its spin; like the ground state, the resultant configuration has a multiplicity of one and is called a singlet state. If the promoted electron reverses its spin (d), the molecule has two electrons of parallel spin and the result is a triplet state with a multiplicity of three. In a closer view, where the repulsive spin-spin interactions between various electrons are regarded, it appears that the energy of the electron in the triplet state is always lower than in the respective singlet state because electrons of like spin tend to avoid occupying the same regions in space. In the energy level diagram of the molecule as a whole (state diagram), from which energy differences corresponding to photon energies can be read, the sum of the energies of all four π electrons has to be taken into account.

After these general considerations on the molecular orbitals, we can now turn to the calculation of the transition energies. In the simplest model, the

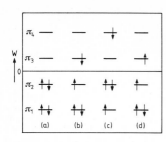

Fig. 3.3. Energy level scheme of some of the π-electron configurations of 1,3 butadiene. (Griffiths 1976). Here α is arbitrarily set zero

free electron molecular orbital (FEMO) method, the assumptions are (a) π electrons can be treated independently of the fixed σ-bonds, (b) in a planar molecule possessing several conjugated double bonds, the π orbitals extend over the whole molecular framework and the electrons are delocalized within the π orbitals, and (c) interelectronic repulsion energies are ignored.

The oscillating potential due to the attractive forces experienced by the electrons as they move past the positively charged nuclei (as indicated in Fig. 3.4) is approximated by a potential $V(x)$ of constant value over the length L of the framework within which the π electrons can move freely, and is limited at the ends by a "wall" where the potential rises to infinity.

The molecular "box" in which the electron is confined can be described quantum-mechanically by Schroedinger's equation in one dimension

$$\frac{\partial^2 \psi}{\partial x^2} + \frac{8\pi^2 m}{h^2} [W - V(x)] \psi = 0 , \tag{3.3}$$

where the average potential energy $\langle V \rangle$ is taken as

$$\langle V \rangle = 0 \quad \text{for} \quad 0 \leqslant x \leqslant L$$
$$\langle V \rangle = \infty \quad \text{for} \quad x < 0 \quad \text{and} \quad x > L . \tag{3.4}$$

Here m is the mass of the electron, h is Planck's constant, and the length of the chain is

$$L = lN , \tag{3.5}$$

where l is the distance between neighbouring C atoms along the chain and N is the number of π electrons. Since the electron is not allowed to leave the box, the waves or wave functions that describe the motion of the electron must go to zero at the ends, or, alternatively stated, the box has to contain an integral number q of half-waves. A direct analogy is the vibration of a string fixed at two points, which occurs only at discrete frequencies. Solutions of Eq. (3.3) corresponding to the boundary conditions for the waves are the wave functions

$$\psi_q = \sqrt{\frac{2}{L}} \sin\left(\frac{\pi x}{L} q\right) , \quad (q = 1, 2, \ldots) \tag{3.6}$$

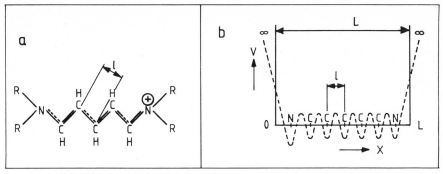

Fig. 3.4. a A linear organic molecule with conjugated double bonds (polymethine dye). Because of delocalization (*dashed lines*) there is effectively one π electron for each internuclear bond. **b** Approximate distribution of the potential energy (*dashed*) and the simplifying assumption (*full lined*) for the FEMO concept. (After Meier 1963)

for $0 \leqslant x \leqslant L$ (Fig. 3.5 a), and inserting Eq. (3.6) into Eq. (3.3) and carrying out the indicated operations results in a unique set of energy eigenvalues

$$W_q = \frac{h^2}{8mL^2} q^2 , \quad (q = 1, 2, \ldots) . \tag{3.7}$$

This model of a one-dimensional electron gas yields the space distribution of the discrete energy states which are possible along the molecular chain. The products $\psi_q \psi_q^*$ are normalized density distributions of the probable positions of the π electron in the different states q (Fig. 3.5 b).

The wave theory of matter independently suggests that the motion of the electron forward and backward across the molecule can be described by a wave whose wavelength λ_B is given by the de Broglie relationship

$$\lambda_B = \frac{h}{mv} , \tag{3.8}$$

where v is the velocity of the electron.

There can only be an integral number of half-wavelengths between the two nodes at the end points, i.e.

$$L = q \frac{\lambda_B}{2} . \tag{3.9}$$

Combining Eqs. (3.8) and (3.9) gives the velocity of the electron as

$$v = \frac{qh}{2mL} . \tag{3.10}$$

The total energy of the electron, i.e. the energy of the molecular orbital, is the sum of the kinetic energy $mv^2/2$ and the potential energy, which was arbitrarily set to zero within the potential well. Hence

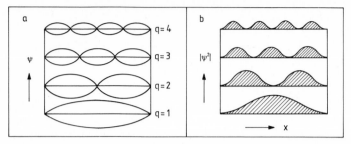

Fig. 3.5. a Eigenfunctions; **b** normalized electron density distributions at the different energy levels. (Meier 1963)

$$W_q = m v^2/2 = \frac{q^2 h^2}{8 m L^2} \ ,$$

confirming that Eq. (3.7) yields the real eigenvalues of the system.

After having determined the possible energy levels, we have to find the actual occupancy of the corresponding molecular orbitals. Filling the orbitals has to obey the Pauli exclusion principle, which means that each orbital starting from the one with the lowest energy can accommodate only two π electrons with antiparallel spin orientations. Thus N delocalized π electrons on a chain of conjugated double bonds such as depicted in Fig. 3.4 can occupy at most $q = N/2$ states in the lowest (ground) energy state of the molecule. The highest occupied orbital represents an energy of $W_{N/2}$, and the lowest empty one has $W_{(N/2)+1}$ (see Fig. 3.6).

Absorption of a photon of proper energy $h\nu$ causes the transition of an electron from orbital q to $q+1$ (from $q = 2$ to $q = 3$ in the example of Fig. 3.6) corresponding to the energy difference

$$\Delta W = W_{(N/2)+1} - W_{N/2} = \frac{h^2}{8 m L^2}(N+1) \ . \tag{3.11}$$

This expression can be modified by recognizing (Fig. 3.4) that $L = Nl$, where l is the internuclear bond length

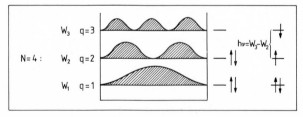

Fig. 3.6. Electron density distribution of a linear chain molecule with four π electrons and occupation of the energy levels in (*left*) the ground state, and (*right*) an excited state

$$\Delta W = \frac{h^2}{8ml^2} \frac{N+1}{N^2} .$$

(3.12)

By equating this with the photon energy $\Delta W = h\nu = hc/\lambda$, we find the longest wavelength λ_{max} which can be absorbed by the molecule due to the first transition

$$\lambda_{max} = \frac{8mc}{h} \frac{N^2 l^2}{N+1} .$$

(3.13)

If the constants are inserted, and we take the mean internuclear bond length of polymethine dye as an example $l = 0.139$ nm (Meier 1963), we arrive at

$$\lambda_{max} = 63.7 \frac{N^2}{N+1} \quad \text{(nm)} .$$

(3.14)

The simple rule derived from Eq. (3.14) is that in the free electron molecular orbital (FEMO) theory, an elongation of the conjugated double bond structure will shift the absorption peak to longer wavelengths. For example, one can introduce additional vinyl groups having a single double bond and two π electrons into the polymethine dye. The corresponding shift of the absorption peak is about 100 nm per double bond. Table 3.1 compares the computed and measured wavelengths of the first transition for increasing numbers of vinyl groups.

In this favourable example the simple model of free electron molecular orbitals works surprisingly well in predicting the main absorption feature. In many other cases, however, the complete separation between σ and π bonds and the perfect delocalization of the π electrons (constant potential over the entire molecular framework), which are the conditions for the FEMO method, are not sufficiently fulfilled. Therefore more refined methods have to be applied, which unfortunately do not allow exact solutions of the Schroedinger equation.

The electron experiences the attractive forces from the positively charged nuclear framework. Therefore a more satisfactory method of obtaining one-electron orbital wave functions and energies is the linear combination of atomic orbitals (LCAO) method. The assumption is made that the molecular orbital wave function ψ_0 can be approximated as a linear combination of the component atomic orbital wave functions ϕ_i.

Table 3.1. Measured and computed (FEMO) wavelengths of the first transition of carbocyanines as a function of the number of electrons. (Meier 1963)

Number of vinyl groups	Number of double bonds	Number of π electrons	λ_{max} (computed) [nm]	λ_{max} (measured) [nm]
0	4	10	576	590
1	5	12	706	710
2	6	14	834	820
3	7	16	959	930

Thus in a system of n overlapping orbitals, the molecular wave function can be written as

$$\psi_0 = c_1\phi_1 + c_2\phi_2 + \ldots + c_n\phi_n . \tag{3.15}$$

The mixing coefficients c_i can have any value between ± 1, and denote the relative contributions of each atomic orbital to ψ_0. Their algebraic signs indicate either in-phase overlap $(+)$ or out-of-phase overlap $(-)$.

The Schroedinger equation for the complete molecular state can, as in Eq. (3.1), be applied to the particular molecular orbital

$$H\psi_0 = W\psi_0 \tag{3.16}$$

to define its eigenvalue or energy W. Multiplying both sides of Eq. (3.16) by the wave function ψ_0^* gives

$$\psi_0^* H\psi_0 = W\psi_0^* \psi_0 , \tag{3.17}$$

because H is an operator and W can be taken as a coefficient. Integrating Eq. (3.17) over all space yields the energy of a molecular orbital, expressed in terms of the atomic orbital wave function and the Hamiltonian operator H

$$W = \frac{\int \psi_0^* H\psi_0 d\xi}{\int |\psi_0|^2 d\xi} = \frac{\int \left[\sum\limits_{i=1}^{n} c_i\phi_i\right]^* H\left[\sum\limits_{j=1}^{n} c_j\phi_j\right] d\xi}{\int \left|\sum\limits_{i=1}^{n} c_i\phi_i\right|^2 d\xi} . \tag{3.18}$$

If an expression (the energy W of the molecular system) contains adjustable parameters (the mixing coefficients c_i), then adjustment of these parameters to give a minimum value of W also gives, according to the variation principle, the best possible value of W, obtainable with the approximation (3.15).

To minimize Eq. (3.18) with respect to the coefficients c_i, it is necessary to differentiate the equation with respect to each coefficient separately and equate the partial derivative to zero, yielding n equations

$$\frac{\partial W}{\partial c_1} = \frac{\partial W}{\partial c_2} = \ldots = \frac{\partial W}{\partial c_n} = 0 . \tag{3.19}$$

The resultant equations are called the secular equations and they contain integrals of the following form and designation

$$\int \phi_i H\phi_i d\xi = H_{ii}$$
$$\int \phi_i H\phi_j d\xi = H_{ij}$$
$$\int \phi_i^2 d\xi = S_{ii} \tag{3.20}$$
$$\int \phi_i\phi_j d\xi = S_{ij} .$$

The individual atomic orbitals ϕ_i are − if regarded separately − subject to the same procedure for determining their respective eigenvalues as indicated in

Eqs. (3.17) and (3.18) for the molecular orbitals. Therefore the expressions (3.20) can be used for normalizing the atomic orbitals and to identify their energy values in a simple way. In the next step of approximations, the Hueckel Molecular Orbital (HMO) method, several assumptions concerning these integrals are made: the S_{ii} integrals can be equated to unity, which normalizes the atomic orbitals. The S_{ij} integrals are a measure of the degree of overlap. These overlap integrals are neglected in the Zero Differential Overlap (ZDO) approximation. In practice, the neglect of overlap integrals does not have a very significant effect on the results, but it simplifies the computation considerably. (Improved molecular orbital models with inclusion of the overlap are reviewed by Peacock 1972.) Each H_{ii} integral is a discrete energy quantity and represents the energy of an electron while it occupies the atomic orbital ϕ_i. Therefore this energy is called the Coulomb integral α_i, and in HMO procedure it is assumed that all H_{ii} integrals for carbon atoms have the same numerical value α. The H_{ij} integrals can be regarded as the energy of an electron while it occupies the region of overlap of orbital ϕ_i with ϕ_j. The H_{ij} integrals are called resonance integrals β_{ij} and in the HMO procedure the resonance integrals of non-adjacent atoms are regarded as zero and the integrals for all adjacent pairs of carbon atoms are assumed to have the same value β.

For a molecule with only two atoms contributing to the relevant molecular wave function as for ethylene ($H_2C = CH_2$), we have the wave function

$$\psi_0 = c_1 \phi_1 + c_2 \phi_2 \tag{3.21}$$

and the corresponding eigenvalues

$$W = \frac{\int (c_1 \phi_1 + c_2 \phi_2) H (c_1 \phi_1 + c_2 \phi_2)\, d\xi}{\int (c_1 \phi_1 + c_2 \phi_2)^2 d\xi} , \tag{3.22}$$

which have to be minimized according to the above procedure. Computation of all products in the numerator and denominator according to Eq. (3.20) yields

$$\frac{c_1^2}{c_1^2 + c_2^2} H_{11} + \frac{c_1 c_2}{c_1^2 + c_2^2} (H_{12} + H_{21}) + \frac{c_2^2}{c_1^2 + c_2^2} H_{22} = W .$$

Applying the above-mentioned assumptions of the Hueckel and ZDO procedures gives

$$\begin{aligned}
&\int \phi_1 H \phi_1 d\xi = \alpha , \quad \int \phi_2 H \phi_2 d\xi = \alpha && \text{Coulomb integrals} \\
&\int \phi_1 H \phi_2 d\xi = \beta , \quad \int \phi_2 H \phi_1 d\xi = \beta && \text{Resonance integrals} \\
&\int \phi_1^2 d\xi = 1 , \quad \int \phi_2^2 d\xi = 1 && \text{Normalized} \\
& && \text{atomic orbitals} \\
&\int \phi_1 \phi_2 d\xi = 0 , \quad \int \phi_2 \phi_1 d\xi = 0 . && \text{Overlap integrals.}
\end{aligned} \tag{3.23}$$

With this the minimization (3.19) can be performed. Partial differentiation of both sides of

$$(c_1^2 + c_2^2)\, \alpha + 2\, c_1 c_2 \beta = (c_1^2 + c_1^2)\, W\ ,$$

with respect to c_1 and c_2 separately, keeping the condition (3.19) in mind, we arrive at

$$c_1(\alpha - W) + c_2\beta = 0$$

$$c_1\beta + c_2(a - W) = 0\ , \tag{3.24}$$

which are the secular equations for this simple molecule. Taking into account that the resonance integrals vanish for all non-adjacent atoms, we easily find the secular equations for a longer molecule of n atomic centres contributing potentials ϕ_i to the molecular orbital as

$$c_1(\alpha - W) + c_2\beta + c_3 0 + \ldots + c_n 0 = 0$$

$$c_1\beta + c_2(\alpha - W) + c_3\beta + \ldots + c_n 0 = 0$$

$$c_1 0 + c_2\beta + c_3(\alpha - W) + \ldots + c_n 0 = 0 \tag{3.25}$$

$$\vdots$$

$$c_1 0 + c_2 0 + \ldots + c_{n-1}\beta + c_n(\alpha - W) = 0\ .$$

From linear algebra it is well known that a homogeneous system of linear equations in the unknown c_i is consistent only if the determinant of the coefficients of these unknown quantities equals zero. The appearance of Eq. (3.25) can be simplified by dividing each term by β and substituting $(W - \alpha)/\beta = x$. The eigenvalues x are then related to the orbital energy by $W = \alpha + x\beta$. The resultant determinant becomes

$$\begin{vmatrix} -x & 1 & 0 & 0 & \ldots & 0 \\ 1 & -x & 1 & 0 & \ldots & 0 \\ 0 & 1 & -x & 1 & \ldots & 0 \\ \vdots & & & & & \\ 0 & 0 & 0 & \ldots & 1 & -x \end{vmatrix} = 0\ . \tag{3.26}$$

Let us regard three examples: Ethylene, 1,3-butadiene and 1,3,5-hexatriene with 2, 4 and 6 atomic centres contributing to the molecular π electron orbitals. The determinant (3.26) yields respectively

$$x^2 - 1 = 0$$

$$x^4 - 3x^2 + 1 = 0$$

$$x^6 - 5x^4 + 6x^2 - 1 = 0\ .$$

The solutions x for the three different molecules are

$$x = \pm 1$$

$$x = \pm 0.618,\ \pm 1.618$$

$$x = \pm 0.445,\ \pm 1.247,\ \pm 1.802\ .$$

The resulting energy diagrams of the π orbitals are presented in Fig. 3.7.

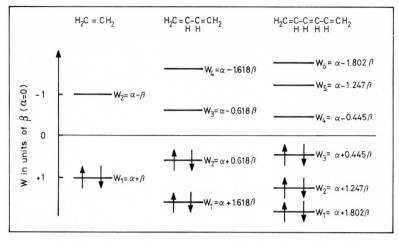

Fig. 3.7. Energy level diagrams of the π orbitals of ethylene, 1,3-butadiene and 1,3,5-hexatriene according to the HMO- and ZDO concepts, in units of β; α is arbitrarily set equal to zero. The occupation by electrons is shown for the ground state

From the knowledge of the eigenvalues for each molecular orbital, the mixing coefficients c_i for each orbital can be evaluated if use is made of the normalization requirement for a molecular orbital. For 1,3-butadiene this reads

$$\int \psi_0^2 d\xi = \int (c_1\phi_1 + c_2\phi_2 + c_3\phi_3 + c_4\phi_4)^2 d\xi = 1 \ . \tag{3.27}$$

Due to the ZDO assumption all crossed terms $\phi_i\phi_j$ for $i \neq j$ are zero, so that Eq. (3.27) reduces to

$$c_1^2 + c_2^2 + c_3^2 + c_4^2 = 1 \ . \tag{3.28}$$

In the case of butadiene, the four secular equations, i.e. the first four lines and first four rows of Eq. (3.25) and the normalization equation (3.28) constitute five equations, from which the four unknown mixing coefficients c_i can be found. These provide the following expressions for the wave functions of the molecular orbitals ψ_0 of 1,3-butadiene (Griffiths 1976) by the LCAO method

$$\psi_1 = 0.371 \ \phi_1 + 0.600 \ \phi_2 + 0.600 \ \phi_3 + 0.371 \ \phi_4$$

$$\psi_2 = 0.600 \ \phi_1 + 0.371 \ \phi_2 - 0.371 \ \phi_3 - 0.600 \ \phi_4$$

$$\psi_3 = 0.600 \ \phi_1 - 0.371 \ \phi_2 - 0.371 \ \phi_3 + 0.600 \ \phi_4 \tag{3.29}$$

$$\psi_4 = 0.371 \ \phi_1 - 0.600 \ \phi_2 + 0.600 \ \phi_3 - 0.371 \ \phi_4 \ .$$

The orbital profiles are drawn in Fig. 3.8, where the size and relative phases of atomic p orbitals reflect the magnitude and sign of each mixing coefficient c_i. The dotted lines indicate the amplitude of the wave functions along the molecular axis. It can be seen that the number of out-of-phase

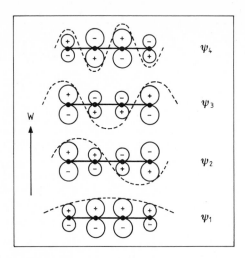

Fig. 3.8. Molecular orbital wave function profiles for 1,3-butadiene. (Griffiths 1976)

overlaps increase with increasing energy of the molecular overlaps. These profiles show a qualitative similarity to the free electron wave functions in Fig. 3.5. Here, however, the interference between the wave functions of the fixed atoms gives rise to the oscillatory shapes of the molecular orbital wave functions and determines the weighting of their respective effects. In contrast to this, the solution (3.7) suggests an infinite number of "string-like" molecular orbitals.

Knowledge of the energy eigenvalues W_j corresponding to the molecular orbital wave functions ψ_j can provide us with the total energy of the π orbital. Filling the energy levels with the available π electrons results in a total electron energy of

$$W_\pi = \sum_{j=1}^{n} b_j W_j = \sum_{j=1}^{n} b_j(\alpha + x_j \beta) \ , \tag{3.30}$$

where the subscript j runs from 1 to n, the number of atomic centres in the π electron system, and b_j is the respective number of π electrons at level j.

Hence for the ground states of the three model molecules ethylene, 1,3-butadiene, 1,3,5-hexatriene we arrive (see also Fig. 3.7) at

$$W_\pi^e = 2(\alpha + \beta) = 2\alpha + 2\beta$$

$$W_\pi^b = 2(\alpha + 0.618\,\beta) + 2(\alpha + 1.618\,\beta) = 4\alpha + 4.472\,\beta \tag{3.31}$$

$$W_\pi^h = 2(\alpha + 0.445\,\beta) + 2(\alpha + 1.247\,\beta) + 2(\alpha + 1.802\,\beta) = 6\alpha + 6.988\,\beta \ ,$$

respectively.

If the π-bond energy is computed under the assumption of localized double bonds, then one can multiply the energy value of ethylene by a factor of 2 to achieve the energy value of 1,3-butadiene, and by a factor of 3 for that of 1,3,5-hexatriene. However, this yields values which are lower by 0.472 and

0.988, respectively, when compared with the HMO model results of Eq. (3.31) computed under the assumption of delocalized double bonds. These differences correspond to the delocalization energy that is gained by conjugation of the two (or three) double bonds.

The transition energy for the promotion of an electron from orbital j to orbital k is given by the difference

$$W_k - W_j = \alpha + x_k\beta - (\alpha + x_j\beta) = (x_k - x_j)\beta \; . \tag{3.32}$$

The HMO method has been successfully applied to a class of molecules showing strong resonance interactions that result in significant bond length equalization along the conjugated chain, thus justifying the constant β approximation. These compounds are the symmetrical cyanine dye molecules like the ones in Table 3.2. The positive charge peculiar to these ions is alternately localized on one or the other of the terminal nitrogen atoms, ensuring a uniform charge density and hence a constant α value for all intermediate carbon atoms.

The computation of the energy levels, electron density distributions and colour properties of more complex ring compounds can also be realized by the procedures described above. The basic assumption in these cases is also the complete delocalization of the π electrons around the whole ring structure, i.e. that single and double bonds are indistinguishable. Because of the missing potential walls in a circular ring system, another boundary condition has to be fulfilled. Only those wave functions are allowed which, after one full cycle around the ring, reproduce themselves (the wave function variation has to contain multiples of a full wavelength over the "circumference" L of the ring). In terms of the free electron molecular orbital (FEMO) theory, this condition can be expressed by

$$L = 2\pi r = q\lambda_B \; , \quad (q = 0, \pm 1, \pm 2 \dots) \; , \tag{3.33}$$

where r is the effective ring "radius". Using the de Broglie wavelength $\lambda_B = h/mv$, the energy levels can again be expressed as kinetic energies $W = mv^2/2$ only if again the potential energy is arbitrarily set zero, and we obtain

$$W_q = \frac{h^2}{2mL^2} q^2 \; . \tag{3.34}$$

Table 3.2. Measured and computed HMO wavelength of the first transition of two cyanine-type molecules. (Griffiths 1976)

Molecule	$\lambda_{max \, (computed)}$ [nm]	$\lambda_{max \, (measured)}$ [nm]
$Me_2N - CH = CH - CH = N^{\oplus}Me_2$	316	309
$Me_2N - (CH = CH)_3 - CH = N^{\oplus}Me_2$	512	511

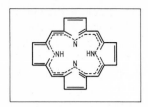

Fig. 3.9. Porphin, the parent structure of the porphyrins. (Weast 1974)

This is equivalent to Eq. (3.7) except for a factor of four that results from the different boundary condition. The quantum number $q = 0$ refers to an even distribution of π-electrons around the ring. The $\pm$ signs, attached to the quantum numbers 1 and higher, signify the twofold degeneracy of the states due to the orbital momentum; in simpler terms the wave functions can be thought of as interfering forward and backward electron waves. The occupation of the states by the π electrons evolves according to the Pauli exclusion principle. For example, in the benzene molecule the six available π electrons occupy the $q = 0$, $q = -1$, $q = +1$ states in pairs with opposite spins. Many examples of ring configurations have been treated with this simple theory and various improved molecular orbital models have been applied (e.g. Peacock 1972).

Of particular interest are macrocycles such as annulenes, which are treated in sizes ranging between 10 and 30 bonds on the ring (Griffiths 1976, and original references cited therein). In connection with remote sensing, chlorophyll is one of the most important of these dyes. It is an example of a class of porphyrins, which are highly coloured heterocyclic compounds that occur widely in nature and are involved in many important biological oxidation, reduction and oxygen transport processes. The parent structure of the porphyrins is porphin ($C_{20}H_{14}N_4$), which is indicated schematically in Fig. 3.9.

The structure contains a 16-bond annulene pathway with maximum delocalization of π electrons (dashed line in the figure).

3.2 Chlorophyll and Spectral Properties of Plants

Chlorophyll is one of the most important biological compounds. It is important not only for plants, but for life on earth as a whole. It acts as a photo-receptor and catalyst for the conversion of sunlight into chemical energy that is necessary for the photochemical synthesis of carbohydrates by plants. During this synthesis of the basic food material for men and animals, oxygen is produced as a by-product. The initial materials for the reaction are simply water and carbon dioxide, the product of respiration, and the decomposition of carbohydrates in the animal organism (respiration) is indeed the inverse process. Hence we can write the reaction as

$$6\,CO_2 + 6\,H_2O \underset{\text{respiration}}{\overset{\text{photosynthesis}}{\rightleftarrows}} C_6H_{12}O_6 + 6\,O_2 . \tag{3.35}$$

A considerable amount of radiation energy is transformed into chemical energy during this process (about 688 kcal per mol) which is stored in the form of adenosine triphosphate (ATP) and becomes available to the plant for the synthesis of carbohydrates. The missing absorptivity of solar light by H_2O and CO_2 is compensated in the cell of the plant by the built-in chlorophylls, which provide the light-absorbing mechanism of the plant. As a consequence of the spectral property needed for this function, chlorophyll has the very pleasant feature of creating the green appearance of the vegetation to the human eye.

Chlorophyll belongs to the class of pyrrole dyes and its structure is related to the porphyrins, as mentioned previously. It exists in two different versions

Chlorophyll a: $C_{55}H_{72}MgN_4O_5$,   and

Chlorphyll b: $C_{55}H_{70}MgN_4O_6$.

Figure 3.10 shows the structures. The natural abundance ratio of the blue-green chlorphyll a and the yellow-green chlorophyll b is approximately $3:1$. The difference between the two structures is only in a single group, (encircled in the figures). Most of the data on the function of the photoactive pigments in the photosynthetic apparatus come from studies of absorption and fluorescence spectra. A typical in vitro absorption spectrum of chlorophyll a and b is shown in Fig. 3.11 (Kamen 1963, with discussion and citation therein).

Two major absorption bands appear for each of the two versions, one at about 430/450 nm, the other at 660/640 nm. To explain the features of the chlorophyll spectra in detail, a more refined approach is needed than the one presented in the previous section. The energy level scheme is complicated considerably by the following fact: Chlorophyll contains a central metal atom (Mg) and an extended conjugated π-system with substituents such as the carbonyl groups. This makes it necessary to consider transitions between the conjugated π-electron system and the metallic atom orbitals, as well as transitions in which electrons move to and from the oxygen and nitrogen atoms. The unshared electrons of the oxygen and nitrogen atoms are thought to be at

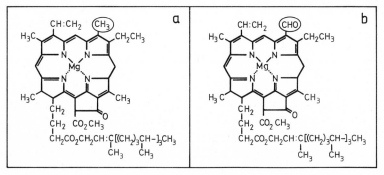

Fig. 3.10. The structural formulas of **a** chlorophyll a and **b** chlorophyll b. (Weast 1974)

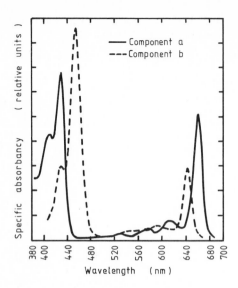

Fig. 3.11. Absorption spectra of chlorophyll a and b (in vitro)

disposal for excitation to the lowest available π orbital. Such transitions are named "n-π" transitions, and often involve energy differences of the same magnitude as the lowest energy π-π transition. For more details the reader is referred to the specialized literature (e.g. Kamen 1963).

The spectral behaviour of chlorophyll is even more complicated by its adaptability. In weak light protochlorophyll is created as a preliminary stage, while under intense light chlorophyll becomes bleached and inactive. However, it is able to regenerate its full activity. If it is partly decomposed, the products of the decomposition become available again for the creation of protochlorophyll. Various reactions of the photosynthesis are described by, e.g., Litvin and Sineshchekov (1975).

The regular arrangement of chlorophyll within the structure of the leaves – almost a crystal-like regularity – permits an intense interaction with the light and a very efficient transport of electrons; the efficiency of the process of photosynthesis of carbohydrate material is estimated to be around 30%.

Investigation on the spectral behaviour of living leaves reveals three different effects which dominate the visible and infrared spectrum.

Figure 3.12 summarizes reflectance, absorbance and transmittance of a mature orange leaf. We can recognize three regions. The wavelength range 0.5 to 0.75 μm is dominated by chlorophyll absorption, there is low reflectance (the leaves appear dark green to the eye) and very low transmittance. The near infrared range 0.75 to 1.35 μm is characterized by high reflectance and transmittance. If the human eye had a sensitivity extending to these wavelengths, nobody would say that leaves are green, but call them infrared. The strong reflectance and transmittance in this spectral region are thought to be caused by internal, regular arrangements of the leaf's constituents and the absence of

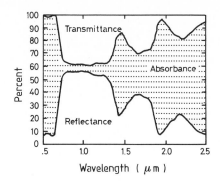

Fig. 3.12. Fractions of the total light incident on the upper surface of a mature orange leaf that are reflected, absorbed and transmitted. (Gausman et al. 1971). Reflectance + Absorbance + Transmittance = 100%

absorption. The ratio of near i.r. and visible reflectance is an indicator of the photosynthesis capacity of the canopy (Sellers 1985). Finally the wavelengths between 1.35 and 2.5 μm are dominated by the water absorption bands at 1.45 and 1.95 μm, thus characterizing the water content of the leaves.

Figure 3.13 demonstrates the effects of different pigments on the visible and near infrared part of the reflectance. The normal chlorophyll-pigmented leaf with its characteristic green reflection peak at around 0.55 μm is compared with anthocyanin pigmentation, in which the characteristic reflection edge at 0.7 μm is shifted to 0.6 μm and the green reflection peak is absent. When both pigments are present, only the green reflection peak is absent, resulting in a purple appearance of the leaf. Without any apparent pigmentation, the leaf shows a high reflectance in the 0.5 to 0.7 μm region and accord-

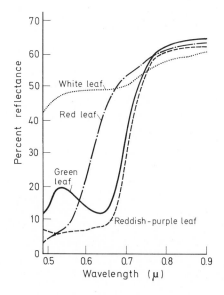

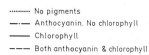

.... No pigments

—·—·— Anthocyanin. No chlorophyll

——— Chlorophyll

– – – Both anthocyanin & chlorophyll

Fig. 3.13. Spectral response in the 0.5 to 0.9 μm region from four pigmentation conditions. (Hoffer and Johannsen 1969)

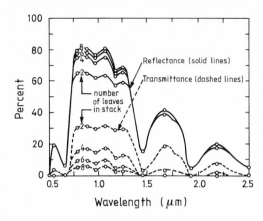

Fig. **3.14.** Reflectance and transmittance of stacked mature cotton leaves. *Lines* represent theoretical curves, *circles* experimental values. (Allen and Richardson 1968)

ingly appears "white". These results show that the colour is indeed a strong indicator of the status of pigmentation.

A convincing demonstration of the transmittance of infrared radiation through leaves is shown in Fig. 3.14.

About 5% of the total light is still transmitted through eight stacked leaves, and even in the reflected radiation a small difference can be recognized between stacks of 6 and 8 leaves.

The water content of a leaf is characterized by the absorptions at 1.45 and 1.95 µm. This is clearly demonstrated by the experiment presented in Fig. 3.15.

Hence the status of pigmentation, the inner structure and the water content of a leaf can be observed almost independently in the three wavelength regions discussed.

Recent developments of airborne imaging high-resolution spectrometers show the ability to distinguish between vegetation species and to identify stress (e.g. Goetz 1985).

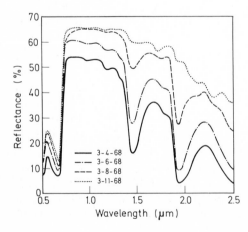

Fig. **3.15.** The effect of progressive leaf drying on spectro-photometrically measured reflectance of upper surfaces of cotton leaves on four dates (March 4 to 11). Each spectrum is the average of 4 leaves. (Myers 1975)

3.3 Polarization of the Media and Dispersion of Radiation

The ability of condensed matter to become polarized by the presence of an electric field is the basis of the dielectric properties of these materials. In other words, the index of refraction is determined by the polarizability of the medium. In Chapter 1 we introduced the permittivity $\varepsilon = \varepsilon_r \varepsilon_0$ and the index of refraction $n = \sqrt{\varepsilon_r}$ as quantities to describe macroscopically the behaviour of the medium under the influence of the electric field of a propagating wave. In this macroscopic picture the polarization P of a medium — in this section we shall concentrate on solids and liquids (condensed matter) — can simply be described by

$$P = \chi \varepsilon_0 E \equiv (\varepsilon_r - 1) \varepsilon_0 E \ , \tag{3.36}$$

where $\chi \equiv \varepsilon_r - 1$ is the electric susceptibility. This macroscopic reaction of the dielectric material to the electric field is caused microscopically by the displacement of charge carriers, which are contained in the material. This means the polarization P is equivalent to the dipole moment per unit volume of the material.

There are various mechanisms which can cause displacement of charge carriers in condensed matter, they perform their action in a different manner and in different time scales.

In each atom of all media, the positively charged nuclei are surrounded by negative electron clouds. If an external electric field is applied, the electrons are displaced slightly with respect to the nuclei. The result is an induced dipole moment, which contributes to the electronic polarization of the material. The electrons are very light and are able to follow the alternating electric field in insulators — almost independent of temperature — up to frequencies of about 10^{15} Hz.

When different atoms form molecules, the electron clouds will normally be displaced eccentrically towards the stronger binding atom. Thus the atoms will not share their electrons symmetrically and therefore acquire charges of opposite polarity, i.e. they behave like ions. An external field will cause a relative displacement of ions of opposite signs inducing another dipole moment. It represents the atomic (or ionic) polarization. This mechanism contributes to polarization at frequencies below those of vibrational modes of molecules (about 10^{13} Hz).

This asymmetric charge distribution between dissimilar atomic partners in a molecule gives rise to a permanent dipole moment, which also exists in the absence of an external electric field. Such dipoles experience a torque in an applied field, which tends to orient them into the field direction. Consequently, an orientational or dipole polarization arises. This contribution to polarization is temperature-dependent and its reaction to alternating fields is much slower (typically at frequencies below those of rotational modes of molecules, i.e. 10^{11} Hz).

In addition to these actions of the electric field on locally bound charges, there are carriers which can migrate for some distance in the structures of solids or liquids, and do this in a preferred direction, when a field is applied. If such carriers are impeded in their motion, either because they are trapped in the potential distribution of the material or at interfaces or, because they cannot be freely discharged at the electrodes, macroscopic space charges and hence field distortions result, indistinguishable from changes in the capacitance, i.e. from changes in the permittivity. This is the space charge or interfacial polarization. It is dependent on the sample size and plays a role only at low frequencies (typically below 10^6 Hz).

Figure 3.16 is a schematic presentation of these four polarization mechanisms (after von Hippel 1954).

We can now try to find a molecular description of the polarization. If each polarizable particle contributes on average a dipole moment $\bar{\mu}$, and if there are N dipole moments per unit volume, then $P = N\bar{\mu}$ is the polarization in units of Ampere seconds per square metre. The polarization is a vector being oriented into the same direction as the average dipole moment and, if the permittivity may be assumed to be a scalar quantity, also the electric field vector is oriented parallel to the polarization. The average dipole moment has approximately a linear dependence on the local electric field E' for small amplitudes of this field $\bar{\mu} = \alpha E'$.

This locally active field E' in general will be different from the externally applied field E. The polarization in the molecular description can thus be formulated as

$$P = N\alpha E' . \tag{3.37}$$

Assuming that the four polarization mechanisms discussed above act independently of each other, we may write the total polarizability α of the medium as the sum of the four respective contributions (dimension As m^2/V)

$$\alpha = \alpha_e + \alpha_a + \alpha_d + \alpha_s \tag{3.38}$$

by the electronic, the atomic, the dipole (orientational) and the space charge polarizabilities. The equality of Eqs. (3.36) and (3.37) yields the link between the macroscopic permittivity and the molecular polarizability.

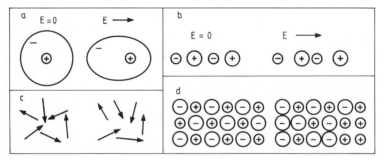

Fig. 3.16. The mechanisms of polarization: **a** electronic, **b** atomic, **c** orientation, **d** space charge polarization

Figure 3.17 is a schematic presentation of the four contributions to the permittivity by the four polarization mechanisms. According to their different dynamical time scales they can follow the alternating electric field, and hence contribute to the polarization only below their respective characteristic frequencies.

Due to the high density of the dipoles in a solid or liquid medium, the action of an applied electric field on a dipole is highly influenced by the effect of the neighbouring dipoles. For the complete link between $\bar{\mu}$ and ε_r we therefore need a relation between the externally applied electric field E and the local field E' acting on the dipole. Such a relation was derived by Mosotti, and it reads

$$E' = E + \frac{P}{3\,\varepsilon_0} = \frac{E}{3}(\varepsilon_r + 2) \ . \tag{3.39}$$

This classical relationship and also more refined concepts are presented by von Hippel (1954), Cole (1961), Böttcher and Bordewijk (1973, 1978). Inserting this local Mosotti field in Eq. (3.37), we obtain the relation between the polarizability per unit volume $N\alpha$ and the relative permittivity ε_r of the material

$$N\alpha = 3\,\varepsilon_0 \frac{\varepsilon_r - 1}{\varepsilon_r + 2} \ . \tag{3.40}$$

Ferro-electric material, for example, is characterized by very high permittivity ($\varepsilon_r \to \infty$) and therefore yields a polarizability of $N\alpha \approx 3\,\varepsilon_0$. For gases at low pressure, $\varepsilon_r - 1 \ll 1$, Eq. (3.40) leads to the simple and evident relation

$$N\alpha \approx \varepsilon_0(\varepsilon_r - 1) \ .$$

A direct link between the average dipole moment $\bar{\mu}$ and the relative permittivity ε_r is easily found by equating (3.36) and (3.37) and using $\bar{\mu} = \alpha E'$,

$$\bar{\mu} = \frac{\varepsilon_0 E}{N}(\varepsilon_r - 1) \tag{3.41}$$

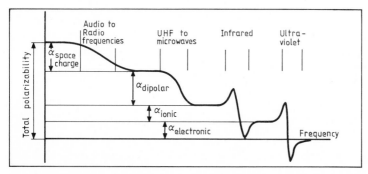

Fig. 3.17. Frequency dependence of the polarizability due to the different contributing mechanisms (schematic)

as long as the linear relationship between dipole moment and the local field holds.

We are dealing here with the interaction of radiation with matter, and this radiation is represented by an electric field oscillating at high frequencies. Therefore we have to give our attention to the dynamic behaviour of the polarization.

In the case of electronic polarization (and to some extent in the case of atomic polarization) the oscillations of the electrons (and ions respectively) caused by the oscillating electric field are only slightly impeded by the structure of the molecule or by neighbouring molecules. Therefore the frequency response can approximately be described by a simple model of point-like masses quasi-elastically bound to an equilibrium position and reacting to field changes like a linear harmonic oscillator. If this oscillator is driven by the force eE', where $E' = E + P/3\,\varepsilon_0$ is the locally acting field in the Mosotti approximation, and where e is the unit electric charge exposed to this field, a dipole moment $\mu = ez$ will be set up due to the instantaneous deviation z of the charge from its equilibrium position. With N oscillators coherently oscillating, a polarization per unit volume

$$P = Nez \qquad (3.42)$$

will be caused.

The dynamic behaviour of an isolated oscillating dipole can be described by a differential equation formally similar to Eq. (1.6), into which we have to include a damping term proportional to the velocity dz/dt of the particle and an external driving force proportional to the locally acting electric field E'

$$m\frac{d^2z}{dt^2} + 2\,\omega_1''m\frac{dz}{dt} + \omega_0^2 mz = eE' \ . \qquad (3.43)$$

Here $2\omega_1''$ is the friction factor, m is the mass of the oscillating particle (electron) or reduced mass if ions of comparable masses are involved, and

$$\omega_0 = \sqrt{f/m} \qquad (3.44)$$

is the resonance angular frequency of the undamped free (without friction and without external forces applied) oscillator, the internal restoring force constant being f [compare with formula (2.16) for the case of molecular vibration frequency]. The friction factor $2\,\omega_1''$ causes the oscillation of the unforced particle to be damped so that

$$z = z_0 \exp\left[i(\omega_0' + i\omega_1'')t\right] \qquad (3.45)$$

describes the free, but damped oscillation, with

$$\omega_0' = \sqrt{\omega_0^2 - \omega_1''^2} \ . \qquad (3.46)$$

the oscillator's resonance frequency, which is lowered with respect to the undamped case.

Making use of Eq. (3.42) and the first part of Eq. (3.39), we can derive, directly from Eq. (3.43), the differential equation of the polarization within a dielectric, consisting of many dipoles per unit volume

$$\frac{d^2P}{dt^2} + 2\omega_1'' \frac{dP}{dt} + \left(\omega_0^2 - \frac{Ne^2}{3\,m\varepsilon_0} \right) P = \frac{Ne^2}{m} E \ . \tag{3.47}$$

The externally applied field E is the driving force and

$$\omega_1' = \sqrt{ \omega_0^2 - \frac{Ne^2}{3\,m\,\varepsilon_0} - 2\,\omega_1''^2 } \tag{3.48}$$

is the angular frequency at which an oscillating electric field $E = E_0 \exp i\omega_1' t$ will cause the polarization to resonate, this resonance being damped by the friction factor $2\omega_1''$. The resonance frequency in this case (an applied harmonic driving force) is lowered with respect to the free undamped, individually oscillating dipole from ω_0 to ω_1' due to polarizing and frictional effects of the surroundings.

The steady-state solution of Eq. (3.47) with an external field $E = E_0 \times \exp(i\omega t)$, harmonically oscillating at some frequency ω, is

$$P = \frac{Ne^2}{m} \frac{E}{\omega_1'^2 - \omega^2 + i2\,\omega_1''\,\omega} \ . \tag{3.49}$$

Because of the complex denominator, which arises when the friction factor $2\omega_1''$ cannot be neglected compared to $(\omega_1'^2 - \omega^2)/\omega$, a phase shift ϕ occurs between the driving field and the resultant polarization, i.e. $P = P_0 \times \exp i(\omega t + \phi)$.

From the macroscopic definition of the polarization (3.36) the relative permittivity can be found as

$$\varepsilon_r = 1 + \frac{P}{\varepsilon_0 E} = 1 + \frac{Ne^2}{\varepsilon_0 m} \frac{1}{\omega_1'^2 - \omega^2 + i2\,\omega_1''\,\omega} \ . \tag{3.50}$$

This formula describes the dispersion of a wave spectrum in a medium according to the concepts of classical physics. The relative permittivity is a complex number, the imaginary part of which is maximum at the frequency $\omega = \omega_1'$, the resonance frequency. The complex permittivity at the resonance frequency is then

$$\varepsilon_r(\omega = \omega_1') = 1 - i \frac{Ne^2}{2\,\varepsilon_0 m \omega_1' \omega_1''} \ . \tag{3.51}$$

At frequencies far below and far above the resonance, the imaginary part becomes negligible. The limiting values of the relative permittivity at $\omega \to 0$ and $\omega \to \infty$ are respectively

$$\varepsilon_r(\omega \to 0) = 1 + \frac{Ne^2}{\varepsilon_0 m \omega_1'^2} \quad \text{and} \quad \varepsilon_r(\omega \to \infty) = 1 \ . \tag{3.52}$$

The widely used notion of plasma frequency, ω_p, can be used here as an abbreviation $\omega_p^2 = Ne^2/\varepsilon_0 m$, hence the imaginary part ε_r'' at resonance and the susceptibility $\varepsilon_r' - 1$ at the low frequency limit can be formulated as

$$\varepsilon_r''(\omega = \omega_1') = \frac{\omega_p^2}{2\,\omega_1'\,\omega_1''} \quad \text{and} \quad \varepsilon_r'(\omega \to 0) - 1 = \frac{\omega_p^2}{\omega_1'^2} \,, \quad \text{respectively .}$$

In a more general physical situation we have to assume that several types of oscillator are present, e.g. different modes of vibration of a molecular bond or different bonds in a molecule (having different charge distributions of the electron clouds and different bond lengths), each exhibiting a different resonant frequency ω_i'. The relative permittivity then contains the sum of the contributions by all types of oscillator which are present in volume densities N_i, having effective masses m_i, and friction constants ω_i'' of each type i

$$\varepsilon_r = 1 + \sum_i \frac{N_i e^2/\varepsilon_0 m_i}{\omega_i'^2 - \omega^2 + i2\,\omega_i''\,\omega} \,, \tag{3.53}$$

when coupling between the different types of oscillators is neglected. Far enough below the lowest resonance frequency, all oscillator types add their characteristic contributions to the static permittivity. If the frequency is increased, each time when crossing the resonance frequency of one type of oscillator, a resonance peak of the imaginary part of ε_r occurs, the strength of it being proportional to the particular number density N_i and inversely proportional to the mass m_i and to the friction constant ω_i''. The real part of ε_r is reduced by the contribution of just the one oscillator type, the resonance frequency of which has been crossed, and only the contributions by the oscillator types exhibiting higher resonance frequencies remain.

Figure 3.18 shows schematically the behaviour of real and imaginary parts of ε_r for the simple case of only one type of oscillator, i.e. the low frequency contribution is only due to one kind of dipole moment. There is only one resonance frequency after which the high frequency value of ε_r' approximately approaches unity because of lack of any other contributions. The spectrum of Fig. 3.18 exactly follows formula (3.50). The maximum value of the imaginary part of ε_r can be read from Eq. (3.51). It is inversely proportional to the friction constant ω_1''. The width of the resonance $\Delta\omega$ is arbitrarily defined by the frequencies at which $\varepsilon_r'' = \varepsilon_{r\max}''/2$. In electronics engineering practice, the full width between these frequencies is taken as the resonance width to compute the quality factor $Q \approx \omega_1'/2\Delta\omega$ of a circuit. A majority of physicists, in particular spectroscopists, take only half of that width for the "half-width" of a spectral resonance line [see also Eq. (2.25)]. This is because $\omega_1 = \omega_1' + i\,\omega_1''$ is a complex frequency which describes the temporal behaviour of a ringing oscillator as we did at Eq. (3.45). In this text we apply the "physicist's concept" of half-width. For reasonably narrow resonances the frequencies for half the absorption maximum are very close to ω_1', the resonance frequency. Therefore the value of $\varepsilon_{r\max}''/2$ is attained approximately at frequencies for which $(\omega_1'^2 - \omega^2)^2 = 4\,\omega_1'^2\,\omega_1''^2$ is fulfilled. Making additional

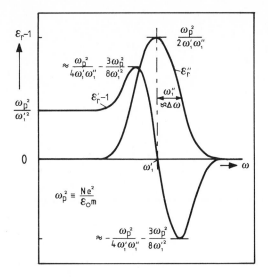

Fig. 3.18. Dispersion and absorption at the resonance frequency. The contribution of one type of oscillators according to Eq. (3.50), $\omega_1'' \ll \omega_1'$ is assumed

use of $\omega_1'' \ll \omega_1'$, we find the half-width of the resonance line, i.e. half the width between the points of half the maximum of the resonance value ε_r'', as

$$\Delta\omega \approx \omega_1'' \ . \tag{3.54}$$

The full spectral width $2\Delta\omega$ is, to a good approximation, identical to the narrow regime of steeply decreasing real part of ε_r. However, since the real part of the permittivity (and that of the refractive index) rises with increasing frequency over the major part of the spectrum, this latter behaviour is called normal dispersion, in contrast to the anomalous dispersion within the half-width region of the spectral line. In the region of normal dispersion the speed of propagating waves decreases for higher frequencies – or in other words – the refraction is stronger for shorter wavelengths.

So far we have studied the dynamic behaviour of polarization which is only slightly impeded by small friction coefficients, and which, therefore, exhibits weakly damped resonances at well-defined frequencies of the oscillating dipole moments. In fact we have formulated oscillation phenomena on a molecular scale by using the concept of classical physics. The results obtained are sufficiently reliable for the purpose of these lectures. In a quantum-mechanical approach the final result of the resonance feature is described in the same way. The half-width of the line can then be identified by either the inverse of the natural lifetime Δt_N of states [Eq. (2.21)], by the Doppler broadened line width (2.25), or by the collision broadened line width

$$\Delta\omega_c = 1/\Delta t_c \ , \tag{3.55}$$

with Δt_c the mean time between collisions.

In order to study the interaction of wave fields with the dynamic properties of dipoles which are strongly disturbed by the presence and motions of neighbouring dipoles – as in the case of orientational polarization – the rapid sequence of interrupting collisions makes an essentially different approach necessary. Because of the extremely short times between collisions ($\Delta t_c \lesssim 1/\omega$) the formation of discrete quantum states is no longer possible. The uncertainty relation, now applied to the energy difference $\Delta W = h\nu$ between quantum states

$$\Delta W \Delta t \geqslant \hbar , \tag{3.56}$$

can no longer be fulfilled, if the time between collisions is shorter than the inverse frequency of the transition $\Delta t_c < 1/\omega_1'$. The resonance state vanishes and the spectral line broadens into a continuum. The classical approach to the treatment of dipole molecules in the condensed phases of liquids and solids, according to Debye (1929), is to consider the polar molecules as rotating in a medium of dominating friction. In terms of our differential equation (3.43), this means that the acceleration term can be neglected.

As one of the most important molecules encountered in remote sensing of the environment, water is constructed in such a way (see Fig. 2.14) that the distribution of the electric charges is asymmetric. Therefore this molecule has a permanent electric dipole moment μ along one axis. Here water may just serve as an – important – example of a polar molecule in the condensed phase. In an electric field forces are exerted on such a dipole, tending to align it along the direction of the field lines. However, the thermal agitation tends to destroy this alignment. These two opposing forces generate a resulting mean dipole moment $\bar{\mu}$ in the direction of the local electric field E'

$$\bar{\mu} = \frac{\mu^2}{3kT} E' , \tag{3.57}$$

as can be found in standard textbooks on solid state physics (e.g. Kittel 1967). The solution of the dynamic behaviour of the mean dipole moment $\bar{\mu}$ (or equivalently of the polarization P) in the case of a harmonic electric field in a friction dominated medium has been derived by Debye (1929) see, e.g. Born (1965) or von Hippel (1954), to be

$$\frac{\mu^2}{3kT} \frac{1}{1+i\omega\tau} . \tag{3.58}$$

Here τ is the relaxation time of this damped process, a phenomenological quantity related to the mean time between collisions Δt_c, and formally linked by $\tau = 1/2\,\omega_1'$ with the friction constant.

Except for the orientation polarization there are also the atomic and electronic polarizations actively contributing to the total polarization. The reaction times of these processes are by orders of magnitude shorter than the period of oscillation relevant to Eq. (3.58). Therefore the induced mean dipole moment at frequencies higher than the relaxation frequency $\omega > 1/\tau$ is determined by

$$\bar{\mu} = (\alpha_e + \alpha_a) E' .$$

Assuming now a local field E' according to Mosotti (3.39), we can formulate the polarizability per unit volume after Eq. (3.40)

$$\frac{N\alpha}{3\,\varepsilon_0} = \frac{N}{3\,\varepsilon_0}\left[\alpha_e + \alpha_a + \frac{\mu^2}{3kT}\,\frac{1}{1+i\omega\tau}\right] = \frac{\varepsilon_r - 1}{\varepsilon_r + 2}\,. \tag{3.59}$$

The limiting values of this expression for $\omega \to 0$ and for $\omega \to \infty$ differ by the term $\mu^2/3kT$ within the bracket.

From Eq. (3.59), the complex permittivity can be formulated (von Hippel 1954) as

$$\varepsilon_r = \varepsilon_{rh} + \frac{\varepsilon_{rs} - \varepsilon_{rh}}{1 + i\omega\tau_e}\,, \tag{3.60}$$

with the static ε_{rs} ($\omega \to 0$) and the high frequency ε_{rh} ($\omega \to \infty$) values, both are assumed to be real quantities and the effective relaxation time constant (Born 1965)

$$\tau_e = \tau\frac{\varepsilon_{rs} + 2}{\varepsilon_{rh} + 2}\,. \tag{3.61}$$

Equation (3.60) is the Debye relaxation formula for the permittivity of a friction-dominated medium in which the internal electric field is assumed to be a Mosotti field. Due to this assumption, the relaxation time is lengthened from τ to τ_e. Equation (3.60) can be represented in real and imaginary parts respectively

$$\varepsilon_r' = \varepsilon_{rh} + \frac{\varepsilon_{rs} - \varepsilon_{rh}}{1 + \omega^2\tau_e^2} \tag{3.62}$$

$$\varepsilon_r'' = \frac{(\varepsilon_{rs} - \varepsilon_{rh})\,\omega\tau_e}{1 + \omega^2\tau_e^2}\,. \tag{3.63}$$

Figure 3.19 shows the relaxation spectrum of liquid water at different temperatures. The real part of the relative permittivity decreases from around 80 at low frequencies (ε_{rs}) to about 5.5 at high frequencies (ε_{rh}). The imaginary part shows its maximum value at the frequency, $\omega = 1/\tau_e$, where the steepest decrease of the real part occurs.

The static contribution to the permittivity is due to the mean dipole moment [Eq. (3.57)] and is therefore dependent on this function of the inverse temperature. In addition to this, the relaxation time constant τ is also inversely proportional to the temperature. This is due to the fact that according to the assumption of dominating friction we may conceive polar molecules as rotating in a liquid of a given viscosity. The friction factor is proportional to this viscosity. In a static applied field the dipoles have a preferential orientation parallel to this field. A sudden removal of this field will cause an exponential decay of this ordered state due to the randomizing agitation of the Brownian movement. The relaxation time measures the time required to

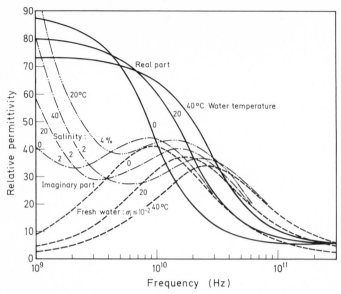

Fig. 3.19. The relaxation spectrum of the orientational polarization of water at various temperatures. The imaginary part is also drawn for various degrees of salinity

reduce the order to $1/e$ of its original value. Under the counteracting influences of a time-dependent electrical field and of the Brownian movements, the relaxation time τ for liquids is found (Born 1965) to be proportional to the viscosity of the liquid, to the effective volume of the regarded polar molecule, and inversely proportional to the temperature. Under the crude assumption of a spherically shaped molecule with radius a, the following approximation:

$$\tau \approx \eta \frac{4\pi a^3}{kT} \tag{3.64}$$

can be used for at least estimating the order of magnitude.

Water at room temperature has a viscosity of $\eta \approx 0.01$ poise and taking in this approximation the molecule as a sphere with an effective radius of $a \approx 1$ Å, a relaxation time constant of $\tau \approx 0.3 \times 10^{-10}$ s can be estimated from Eq. (3.64).

Figure 3.20 presents the temperature dependence of τ_e and of ε_{rs} as determined experimentally. The values of $2\pi\tau_e$ are given to simplify computing the relaxation frequency $1/2\pi\tau_e$ in Hertz. From the figure it becomes obvious that the experimentally determined τ_e is more than one order of magnitude shorter than the estimation (3.64) would yield if one takes the bond length between the O- and the H-atoms as the radius of the "spherical" molecule. When maintaining the spherical model one has to assume a radius of about 0.3 Å in order to match the observation. This reduction of the extent

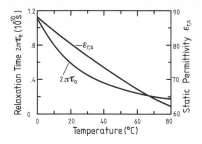

Fig. 3.20. Relaxation time and static value of the relative permittivity of water as a function of temperature. (After Hasted 1961)

of the sphere takes into account an effective occupation of only about 3% of the volume of the sphere by the two hydrogen atoms. Other reasons for the discrepancy between the observed and computed τ_e-values can be: the concept of Mosotti field is not applicable and Eq. (3.61) does not correctly relate $\tau_e(\tau)$. A discussion of the relaxation behaviour of water was given by Hasted (1974).

An additional factor can become very important in the dielectric properties of water: the salinity. A small amount of salt leads to considerable ionic conduction, which is very effective at the lower frequency end of the microwave spectrum. In the simplest description we can ascribe a single valued conductivity σ_i to this ionic contribution. The relation between the conductivity of a medium and the imaginary part of the permittivity,

$$\varepsilon''_{r\,\mathrm{ion}} = \frac{\sigma_i}{\omega\varepsilon_0} \, , \tag{3.65}$$

shows the importance of this effect at the lower frequencies. Figure 3.19 indicates the effect of 2% and 4% salinity on the imaginary part of the permittivity.

In Chapter 1.3 we have seen how the imaginary part of the permittivity is reflected in the penetration depth of the electromagnetic waves. Starting at Eqs. (1.54) through (1.59), the penetration depth can be determined. If we do so for pure water, assuming only a single dispersion maximum of ε''_r due to the relaxation in the microwave region, we arrive at the curve d_I in Fig. 3.21, the penetration depth of the intensity.

From the approximate relation (1.61), valid sufficiently far away from the relaxation frequency ($\varepsilon''_r \ll \varepsilon_r$), and written in an alternate form

$$d_I \approx \frac{\lambda_0}{2\pi} \frac{\sqrt{\varepsilon'_r}}{\varepsilon''_r} \, , \tag{3.66}$$

it can be seen that the penetration depth decreases with increasing frequency. This is not only due to a high loss term ε''_r at the relaxation, but also to the decreasing product of wavelength and real part of the permittivity. Obviously both decrease with frequency around an anomalous dispersion. One has to go very far beyond the relaxation frequency, where ε''_r attains extremely small values, and additionally one needs a normal dispersion, with increasing real

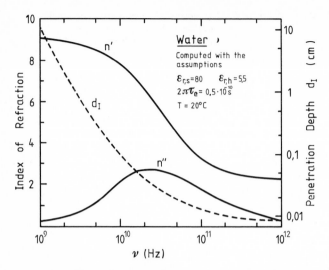

Fig. 3.21. The real (n') and imaginary (n'') parts of the complex index of refraction of water and the penetration depth d_I of the wave intensity

part ε_r' in order to arrive at an increasing penetration depth. The decrease of ε_r'' far beyond the relaxation frequency $1/\tau_e$ goes as

$$\varepsilon_r''(\omega \gg 1/\tau_e) \approx \frac{\varepsilon_{rs} - \varepsilon_{rh}}{\omega \tau_e} , \qquad (3.67)$$

according to the Debye formula (3.63) and therefore cancels the decrease of λ_0. In Fig. 3.21, besides d_I, also n' and n'' related to ε_r by Eq. (1.54) are drawn over the frequency range covering the whole Debye relaxation region of pure water.

In the visible spectrum the concept of dipole relaxation is no longer valid, and we enter the frequency region of a polarization mechanism behaving according to Eq. (3.47). This exhibits a broad spectral region of normal dispersion below the resonance described by Eq. (3.50); a part of it is presented in Fig. 3.18.

Figure 3.22 presents the increase of the index of refraction of water between 0.5×10^{15} Hz and 1.2×10^{15} Hz, i.e. below the resonance region of the electronic polarization. We could ask whether this change of n over the visible spectrum has some effect on the colour of water. We apply the first of Eqs. (1.51) with $n = 1$ and $n_2 = n_{\text{water}}$ which changes from, say, 1.37 to 1.33 between blue and red. This yields a change of r_r between 0.024 and 0.020, insufficient to explain the blue appearance of water!

Concerning the penetration depth of visible light into water, one can use a very simplified consideration in order to achieve at least a rough estimate. From Eq. (3.50) we can derive that for ω sufficiently far below ω_1', the resonance frequency (normal dispersion), the penetration depth is approximately

$$d_{I\text{res}}(\omega \ll \omega_1') \approx \frac{c}{2\,\omega_1''} \left(\frac{\omega_1'}{\omega}\right)^2 \frac{\sqrt{\varepsilon_{ri}}}{\varepsilon_{ri} - 1} , \qquad (3.68)$$

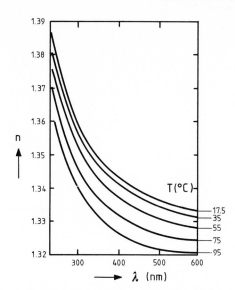

with $\varepsilon_{ri} \equiv \varepsilon_r(\omega \to 0)$ according to (3.52), where the limiting value of the permittivity below the resonance [first equation of (3.52)] is set equal to the high frequency value of the relaxation (ε_{rh}). The penetration depth at frequencies above the resonance can approximately be found as

$$d_{I\text{res}}(\omega \gg \omega_1') \approx \frac{c}{2\omega_1''}\left(\frac{\omega}{\omega_1'}\right)^2 \frac{1}{\varepsilon_{ri}-1}. \tag{3.69}$$

These can be compared with the value achieved from Eq. (3.67) at frequencies sufficiently above the relaxation frequency ($\omega \gg 1/\tau_e$), but sufficiently below the resonance ($\omega \ll \omega_1'$)

$$d_{I\text{rel}}(\omega \gg \omega_e) \approx \frac{c}{\omega_e} \frac{\sqrt{\varepsilon_{rs}}}{\varepsilon_{rs}-\varepsilon_{rh}}, \tag{3.70}$$

where $\omega_e \equiv 1/\tau_e$ is used.

In a crude comparison, the ratio of Eqs. (3.68) and (3.70) for $\omega_e \ll \omega \ll \omega_1'$ and $\varepsilon_{rs} \gg \varepsilon_{rh} \approx 5$ can be approximated by

$$\frac{d_{I\text{res}}}{d_{I\text{rel}}} \approx \frac{1}{2}\frac{\varepsilon_{rs}-\varepsilon_{rh}}{\varepsilon_{ri}-1}\sqrt{\frac{\varepsilon_{ri}}{\varepsilon_{rh}}}\frac{\omega_e}{\omega_1''}\left(\frac{\omega_1'}{\omega}\right)^2. \tag{3.71}$$

If the relaxation frequency ω_e and the halfwidth of the resonance ω_1'' are assumed to be within the same order of magnitude, we can conclude from Eq. (3.71) that the penetration depth of visible light may be several orders of magnitude larger when explained by the resonance type of dispersion (3.50) as it would result from a relaxation model (3.60).

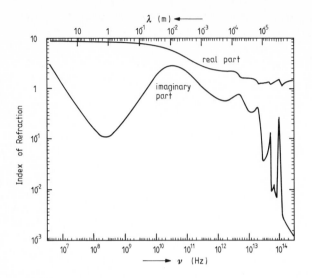

Fig. 3.23. The complex index of refraction of water at 25 °C from radio to infrared frequencies due to an analytic model and the verification by many measured data points. (After Ray 1972)

In Fig. 3.23, for the sake of completeness the complex index of refraction over the dispersion regions of water occurring between about 3 MHz ($\lambda = 100$ m) and 300 THz ($\lambda = 1$ μm) are summarized after Ray (1972), who compiled the measured data from many sources and compared them with a model computation. Improved data in the range 300 MHz to 100 THz have been presented by Hasted (1974). Figure 3.24 shows the index of refraction of ice at 0 °C and −20 °C over an even much wider spectral range from 30 Hz ($\lambda = 10^4$ km) to 300 THz.

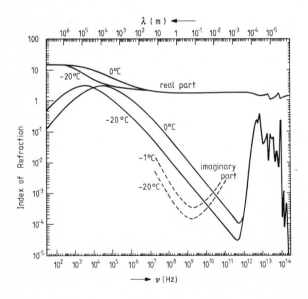

Fig. 3.24. The index of refraction of ice at 0 °C and −20 °C from audio to infrared frequencies according to an analytic model and verification by measurements. (After Ray 1972). Corrections of the imaginary part are included due to recent data by Maetzler (personal communication 1985)

The mobilities of the charge carriers in both solid and liquid state of the same chemical compound are so different that the polarization mechanisms change extremely. Within the range from 10^4 Hz to almost 10^{10} Hz the dielectric constants are either about 80 (water) or 3.2 (ice), and the relaxation frequency is shifted from the 10 GHz to the 10 kHz range correspondingly.

4 Scattering of Radiation

4.1 Light Scattering by Molecules

The absorption of radiation by a molecule and the consequent change of the molecule's dynamic energy by a discrete amount, as discussed in connection with spectral lines in Chapters 2 and in 3.1, is a resonant and inelastic scattering process: resonant because the photon energy has to fit exactly the energy jump $W_n - W_m$ of the molecule from a lower to a higher value, inelastic because the energy of the photon is completely transformed into energy of the molecule. Scattering mechanisms also exist which are non-resonant and/or elastic, but most of them are much less effective than the resonant absorption process. Some of them will now be described in this and the next section.

We know that the electric field E of radiation may induce a dipole moment

$$\mu = \alpha E \tag{4.1}$$

in a molecule, e.g. by electronic polarization (see Chap. 3.3). If a dipole moment is oscillating along a given direction, it behaves like a tiny linear antenna (also called a Hertzian dipole) and radiates electromagnetic waves in a frequency range determined by the dynamic behaviour of the dipole moment. According to classical electrodynamics (e.g. Sommerfeld 1948) the intensity of this radiation at a distance R from the dipole is given by

$$I = \frac{1}{16\,\pi^2 \varepsilon_0 c^3} \frac{\sin^2 \Gamma}{R^2} \left(\frac{d^2\mu}{dt^2}\right)^2 . \tag{4.2}$$

Here Γ is the angle between the direction of the dipole axis and the direction of observation and $d^2\mu/dt^2$ is the acceleration of the electric charge distribution which constitutes the dipole. This latter term represents the dynamic behaviour of the dipole. The intensity decreases obviously as the inverse squared distance. From Eq. (4.2) the total power radiated by the oscillating dipole follows from integration over the surface of a sphere with radius R as

$$P_s = \frac{1}{6\,\pi \varepsilon_0 c^3} \left(\frac{d^2\mu}{dt^2}\right)^2 . \tag{4.3}$$

A harmonically oscillating dipole moment created by a sinusoidal wave field,

$$\mu = \alpha E_0 \sin \omega t , \tag{4.4}$$

results in a radiated intensity, averaged over time, as a function of the angle of observation Γ

$$I(\Gamma) = \frac{c\pi^2 \sin^2\Gamma}{2\,\varepsilon_0 R^2 \lambda^4}\, \alpha^2 E_0^2 \ . \tag{4.5}$$

The total power becomes for this case

$$P_s = \frac{4c\pi^3}{3\,\varepsilon_0 \lambda^4}\, \alpha^2 E_0^2 \ . \tag{4.6}$$

Equations (4.5) and (4.6) are the general formulation of Rayleigh scattering for a single dipole.

Two important conclusions can be drawn from Eqs. (4.5) and (4.6). The intensity and the power of the scattered wave are proportional to λ^{-4}. This strong wavelength dependence explains the blue appearance of the sky on sunny days. The shorter wavelengths are scattered much more efficiently than the longer ones. Secondly, there is no scattered radiation along the axis of the dipole. As regards a single charge oscillating with the electric field of the incident wave, this means that in the polarization direction of this wave no scattering takes place. However, natural light is usually unpolarized, so that this effect is not noticeable in the blue sky.

Let us now consider the dynamic behaviour of the dipole moment more closely. In the case of a charge e with mass m bound to a molecule, a restoring force (spring constant f) is caused when forced to oscillate with the frequency ω of the wave field $E = E_0 \sin \omega t$. We can formulate the dynamic behaviour as we did in Eq. (3.43). Assuming a very dilute medium (atmosphere), the locally active field E' may be approximated by the vacuum field E and the friction may be neglected. The dipole moment is

$$\mu = ez \ , \tag{4.7}$$

with z the instantaneous deviation of the electron from its equilibrium position, and the differential equation for the dipole moment becomes

$$\frac{d^2\mu}{dt^2} + \omega_0^2 \mu = \frac{e^2 E_0}{m} \sin \omega t \ , \tag{4.8}$$

with $\omega_0 = \sqrt{f/m}$. A solution of Eq. (4.8) is

$$\mu = \frac{e^2 E_0 \sin \omega t}{m(\omega_0^2 - \omega^2)} \ , \tag{4.9}$$

and by comparison with Eq. (4.1) it follows that the polarizability for a single charge is

$$\alpha = \frac{e^2}{m(\omega_0^2 - \omega^2)} \ . \tag{4.10}$$

This can be compared with Eq. (3.49), where the polarization P for N charges per unit volume and with damped oscillation has been derived. With

Eqs. (4.10) in (4.5) and (4.6) the dependence of the intensity on the angle Γ between the directions of polarization and of observation and the total power scattered by a single dipole can be determined.

The effectiveness of a scatterer can be quantified by the scattering cross-section. The directional (differential) scattering cross-section $\sigma(\Gamma)$ is defined as the ratio of the scattered power per unit solid angle into a given direction, $I(\Gamma)R^2$, and the intensity of the incoming plane wave $\dfrac{c\,\varepsilon_0}{2}E_0^2$. (In the literature the differential cross-section is often designated by $d\sigma/d\Omega$.)

Hence

$$\sigma(\Gamma) = \frac{\alpha^2\pi^2}{\varepsilon_0^2\lambda^4}\sin^2\Gamma \; , \tag{4.11}$$

having the dimension of square metres per unit solid angle. The total scattering cross-section [$\sigma(\Gamma)$ integrated over all angles surrounding the particle] is the ratio of the total scattered power and the incoming intensity

$$\sigma_s = \frac{8\,\pi^3 a^2}{3\,\varepsilon_0^2\lambda^4} \; . \tag{4.12}$$

For a weakly damped oscillating dipole, the polarizability (4.10) can approximately be applied. In the neighbourhood of the resonance frequency ω_0 the scattering cross-section can grow to extremely high values before it becomes limited by a weak damping. This resonant elastic scattering is therefore much more efficient than the non-resonant elastic (Rayleigh) scattering.

So far we have discussed elastic scattering of radiation, i.e. neither the radiation nor the particle by which the radiation is scattered (e.g. a molecule) have changed their respective energies. If, for example, a molecule is vibrating at its proper vibration frequency ω_v (see Chap. 2), then the ability of the molecule to become polarized, the polarizability α, is oscillating with the vibration frequency about its mean value α_0. Assuming a weak modulation ($\alpha_1 \ll \alpha_0$), the resulting polarizability

$$\alpha = \alpha_0 + \alpha_1 \sin\omega_v t \tag{4.13}$$

will cause the polarization induced by the incident wave field at frequency ω to be modulated by the vibration as

$$\mu = \alpha_0 E_0 \sin\omega t + \tfrac{1}{2}\alpha_1 E_0 [\cos(\omega-\omega_v)t - \cos(\omega+\omega_v)t] \; . \tag{4.14}$$

The first term is the Rayleigh (elastic) term. The second and third term represent the Stokes and Anti-Stokes lines of the Raman scattering respectively; they are the inelastic side bands of the Rayleigh scattering. From the point of view of quantum mechanics, Raman scattering can be explained by the energy levels of the vibrational modes of a molecule. The incident photon of energy $\hbar\omega$ excites the molecule to a very short-lived ($\leq 10^{14}$ s) virtual level at an energy $\hbar\omega$ above the original vibrational level – say W_m. For the decay the molecule can either return to its original level W_m or to a neighbouring level W_n which can correspond, say, to a higher vibrational state differing from W_m by $\hbar\omega_v$. By this process the molecule has gained the energy $\hbar\omega_v$, while the radiation has lost this energy amount. On the other hand, a molecule at energy W_n hit by a photon can be excited to $W_n + \hbar\omega$, a virtual level, and can either decay back to W_n or one of the neighbouring levels, e.g. to the lower vibrational W_m. Because $W_n - W_m = \hbar\omega_v$, after this scattering process the molecule has lowered its energy by the amount $\hbar\omega_v$, which has been transferred to the radiation, leaving this process with energy $\hbar(\omega + \omega_v)$.

Figure 4.1 is a schematic presentation of these three components of the scattering process. There are, of course, many vibrational levels and the energy of the incident photon may combine with any of the vibrational energy levels. For rotating molecules an expression similar to Eq. (4.14) is obtained. However, the Raman shifts are $2\omega_r$ because of the rotational symmetry of the molecule (any polarization situation is identically present twice every revolution). One observes that the Stokes lines are always more intense than the Anti-Stokes lines. Due to the Boltzmann distribution function, the higher population of the lower levels gives dominance to that process for which more potential participants are available. Due to the quantum-mechanical selection rules for vibrations $\Delta v = \pm 1$ and for rotations $\Delta J = 0, \pm 2$, in vibrational-rotational Raman scattering the spectrum is given (Svanberg 1978) by

$$
\nu = \nu_v
\begin{cases}
+\dfrac{4B}{h}\left(J+\dfrac{3}{2}\right) & \text{for } J \to J+2 \\[2mm]
\pm 0 & \text{for } J \to J \\[2mm]
-\dfrac{4B}{h}\left(J-\dfrac{1}{2}\right) & \text{for } J \to J-2
\end{cases}
\tag{4.15}
$$

with the terminology of Chapter 2.1.

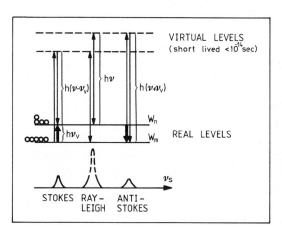

Fig. 4.1. Energy level diagram for Raman scattering. The asymmetry between the intensities of the Stokes- and Anti-Stokes lines is based on the Boltzmann distributed occupancy of the real levels. Here ν_s is the frequency of the scattered radiation. The notations $h\nu = \hbar\omega$ for the photon energy are equivalent

Regarding the photon energies available in light scattering experiments using lidar techniques, one has also to take into account the possibility of electronic transitions and fluorescence. Lasers can provide useful amounts of energy at discrete, determinable wavelengths and one can therefore make practical use of various inelastic scattering processes.

Figure 4.2 shows some examples of such processes which can best be explained with the aid of an energy level diagram (Collis and Russell 1976a). The bands of closely spaced energy levels consist of different vibrational and rotational (not specified here) states all with the same electronic state. The large gaps between the different bands correspond to a change of the electronic state of one of the molecule's constituent atoms. Each species of molecule has a characteristic energy level scheme (and corresponding spectrum) which provides a means of species identification. In the diagrams of Fig. 4.2, some vibrational levels of the electronic ground state and of an excited electronic state are shown, $W_L = h\nu_L$ and $W_S = h\nu_s$ are the energies of the incident (Laser) and of the scattered photons respectively.

Case (a) is ordinary Raman scattering (ORS) as already discussed. W_L is much less than the energy gap between the electronic ground and excited states. The case where the molecule returns from the virtual state to its initial state $(W_S = W_L)$ is merely the elastic Rayleigh scattering. If $W_S \neq W_L$ (Ordinary Raman Scattering) then the energy difference $W_L - W_S$ is characteristic of the species of the scattering molecule (e.g. the energy difference between neighbouring vibrational levels). Therefore the measurement of this energy difference (frequency) in a scattering experiment provides a means of species identification. The incident photon energy (laser frequency) is not restricted to specific values determined by the energy levels of the scattering molecule, thus the Raman scattering process is in principle of broad applicability. However, it is a rather inefficient scattering process because the probability of the excited molecule decaying to any given final state different from the initial one is considerably less than the probability for $W_S = W_L$ (Rayleigh scattering). Measurable return signals in remote atmospheric probing by the Raman effect are obtainable at large distances only for the most abundant gases (e.g. N_2, O_2, H_2O).

If the energy of the incident photon is increased (case b), the scattering cross-section will increase as the excited molecule energy approaches that of the lowest vibrational state in the excited electronic band. Such a process is called resonance Raman scattering (RRS). The increase in effectiveness of RRS over ORS has to be traded against the disadvantage in that the incident photon energy (hence laser frequency) must be chosen to fall near the energy level difference of the specific type of molecule it is desired to detect. In the case (c) of Fig. 4.2, the excited molecule energy falls between the vibrational levels of the excited electronic band, but still outside the width of any vibrational level. This process is called off-resonance fluorescence (ORF). For a given molecular band, the scattering cross-section for ORF is slightly larger than that for RRS. Case (d) of Fig. 4.2 is resonance fluorescence (RF), where

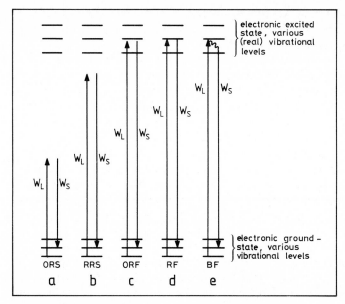

Fig. 4.2. Energy level schemes of various inelastic scattering processes

the excited molecule energy falls within the width of a vibrational level. For a given molecular band, the cross-section for RF exceeds that of any of the processes described so far. The disadvantage, however, is that the laser frequency has to be tuned very carefully to exactly the energy level difference for excitation. If the excited molecule returns to its initial state, the process is elastic resonant scattering, i.e. after a short life-time (typically 10^{-14} to 10^{-6} s) the absorbed energy is re-emitted. As already shown in connection with formulae (4.9) through (4.12), in elastic resonant scattering the cross-section greatly exceeds that of ordinary (elastic) Rayleigh scattering, thus permitting much greater sensitivity in detecting trace gases.

If the energy of the excited molecule falls within the excited electronic band, it may happen to collide with another molecule and in the collision process make a transition to a new intermediate state before it can re-emit a photon [case (e) in Fig. 4.2]. The energy of the new intermediate state may actually fall between the lines shown on the diagram according to the energy lost to molecular translation, and a new rotational level may be entered. Therefore W_S can take on a nearly continuous range of values, so that this process gives rise to a broad continuum of photon energies in the scattered spectrum. In this process, called broad fluorescence (BF), the scattering cross-section of a given molecule in general assumes greater values than for RF, and is hence greater than for any of the other inelastic scattering processes.

Table 4.1 summarizes the sensitivities and life-times of the elastic and inelastic scattering processes discussed here, and for comparison (resonant)

Table 4.1. Typical cross-sections and radiative life-times of various elastic and inelastic interactions. (After Collis and Russell 1976a)

	Process	Differential cross-section (cm^2 sterad^{-1})	Radiative life-times (s)
elastic	Rayleigh scattering	10^{-27}	$\leq 10^{-14}$
	Resonance (elastic) scattering	10^{-27} to 10^{-20}	10^{-14} to 10^{-6}
	Mie scattering	10^{-27} to 10^{-9}	$\leq 10^{-12}$ (depends on size of scatterer)
inelastic	Ordinary Raman scattering	10^{-30} to 10^{-29}	$\leq 10^{-14}$
	Resonance Raman scattering	10^{-30} to 10^{-23}	10^{-14} to 10^{-8}
	Off-resonance fluorescence	$\sim 10^{-23}$	10^{-10} to 10^{-8}
	Resonance fluorescence	10^{-23} to 10^{-16}	10^{-8} to 10^{-1}
	Broad fluorescence	$\sim 10^{-16}$	$\sim 10^{-8}$
	Resonant absorption	Total cross-section $\sim 10^{-15}$ cm^2	

absorption and Mie scattering are included. The latter will be discussed in the next section and is not to be regarded as a molecular scattering process.

The numbers in Table 4.1 give only an indication of the relative strengths, individual processes showing a wide range of values. The strongest σ-values for fluorescence only occur if the pressure is low as in the stratosphere, otherwise radiationless transitions caused by collisions reduce σ by a factor down to 10^{-5}. The radiative life-times are important, because they determine the accuracy with which target ranges may be determined.

As an example of the detection of atmospheric constituents by Raman scattering, Fig. 4.3 shows the Raman spectrum of an oil smokeplume. For this experiment a pulsed nitrogen (N_2) laser was used producing 20 kW peak power of 10 ns pulse length with a repetition rate of 50 pps at the wavelength of 337.1 nm. The resulting spectra, obtained with a single-grating monochromator, are shown in Fig. 4.3. The small cross-sections of Raman scattering severely limit its ability to detect pollutant gas at typical atmospheric concentrations (several parts per million) from great distances.

Table 4.2 indicates for several selected molecules strong lines at shifted wavelengths λ_R due to Raman scattering, assuming an incident wavelength for $\lambda_L = 337.1$ nm of the pulsed nitrogen laser.

An example of resonant elastic scattering has been reported by Hake et al. (1972). Because the effectiveness of resonance scattering is strongly reduced in the troposphere, it has been mainly applied to observations of the middle atmosphere, particularly successfully in the observation of the layer of free atomic sodium near 90 km.

The reported results (Fig. 4.4) have been obtained by night-time measurements. An interesting result was the observation of the enhancement of resonantly scattered laser returns at the time of the Geminids meteor shower,

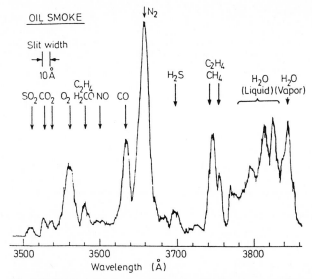

Fig. 4.3. Spectral distribution of Raman-shifted components in an oil smokeplume. The spectra were obtained from a distance of a few hundred metres using a pulsed nitrogen laser ($\lambda_L = 337.1$ nm) in a Raman-lidar experiment. (Inaba and Kobayashi 1972)

Table 4.2. Shifted wavelengths due to Raman scattering of a laser line at 337.1 nm. (After Collis and Russel 1976a)

Molecule	N_2	O_2	CO_2	SO_2	CO	CH_4	H_2S	H_2O
λ_R (nm)	365.8	355.9	353.7	350.8	363.5	373.9	369.7	384.4

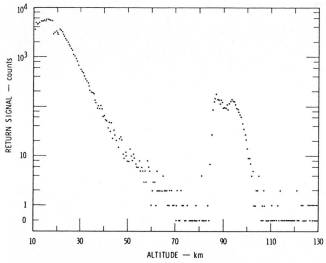

Fig. 4.4. Return signal from a 10-min observation of the 90-km atomic sodium layer using a ground-based tuneable dye laser emitting at 589 nm. (Hake et al. 1972)

in December 1971. These results were interpreted as evidence for meteor showers as one source of the 90-km sodium layer.

The fluorescence effects offer a great potential for remote sensing of various species, e.g. dissolved organic carbon and chlorophyll concentrations in water (Bristow et al. 1985).

4.2 Scattering of Radiation by Macroscopic Particles

If many molecules cluster to form liquid or solid particles, the character of the scattering is changed. Also the dimension of such clusters can grow to values no longer negligible compared with the wavelength ranges used for remote sensing purposes, and perhaps even larger than the wavelength.

Figure 4.5 shows approximate size ranges of aerosols and hydrometeors. Aerosols are particulate matter which can remain suspended in the atmosphere for long periods, of the order of days. Examples include smog, smoke, haze; their sizes are generally under 1 µm in radius. Dust particles have sizes up to about 0.1 mm. Hydrometeors are water particles in liquid or solid form, examples being fog, mist, rain, snow, and hail, the sizes of which generally range from more than 1 µm to about 1 cm.

In several cases, the shape of the scatterers can be regarded as spherical, but in some (snow flakes) a sphere would be a very crude approximation of the actual shape. As long as the characteristic dimension (in the case of a sphere, for example, the circumference) is much smaller than the wavelength, the scattering particle may be regarded as a small (Hertzian) dipole and the concept of Rayleigh scattering can well be applied. As soon as the dimensions become comparable to the wavelength, the scattering must be regarded as the result of interfering waves which partly pass around the particle and partly penetrate through the particle if it is dielectric. This type of interaction is called Mie scattering. If the particle dimensions are much larger than the wavelength, the solution of the scattering problem reduces to geometric optics, i.e. the scattering cross-section approaches the geometric cross-section of the particle.

In the case of small particles ($2\pi a \ll \lambda$) we may follow the concept of Rayleigh scattering by a small dipole, as expressed by formulae (4.5) and (4.6)

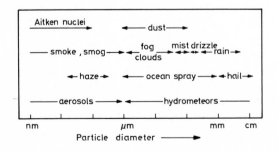

Fig. 4.5. Approximate size ranges of various particles in the atmosphere. (Ishimaru 1978)

for the angular distribution of the scattered intensity and for the total scattered power respectively. However, in the present case of a particle consisting of many molecules, it is better to adopt a macroscopic description of the polarizability. In Chapter 3.3 we discussed the polarization of dielectric media consisting of N dipoles per cubic metre and in Eq. (3.41) found a value for $N\bar{\mu}$ when the Mosotti field was assumed as the locally active electric field. With this approach, the average dipole moment $\bar{\mu}$ used instead of μ in Eq. (4.2) and (4.3), we arrive at

$$I(\Gamma) = \frac{9c\varepsilon_0\pi^2}{2R^2\lambda^4N^2}\left(\frac{\varepsilon_r-1}{\varepsilon_r+2}\right)^2 E_0^2\sin^2\Gamma \tag{4.16}$$

for the scattered intensity per average dipole as a function of the angle Γ between the direction of the electric dipole axis (direction of polarization) and the direction of propagation of the scattered wave (direction of observation). The total scattered power becomes

$$P_s = \frac{12c\varepsilon_0\pi^3}{\lambda^4N^2}\left(\frac{\varepsilon_r-1}{\varepsilon_r+2}\right)^2 E_0^2. \tag{4.17}$$

The notations are as previously introduced. The effective cross-section for scattering has been defined for Eq. (4.11) and (4.12). Consequently the directional (differential) cross-section of a single particle can now be defined by

$$\sigma(\Gamma) \equiv 2\frac{I(\Gamma)R^2}{c\varepsilon_0E_0^2} = \frac{9\pi^2}{\lambda^4N^2}\left(\frac{\varepsilon_r-1}{\varepsilon_r+2}\right)^2\sin^2\Gamma, \tag{4.18}$$

and the total scattering cross-section (unit m^2) per dipole

$$\sigma_s \equiv 2\frac{P_s}{c\varepsilon_0E_0^2} = \frac{24\pi^3}{\lambda^4N^2}\left(\frac{\varepsilon_r-1}{\varepsilon_r+2}\right)^2. \tag{4.19}$$

As a new concept we now introduce the Rayleigh scattering function for volume scattering as the resulting differential scattering cross-section caused by all the N particles contained in the unit volume

$$\gamma(\lambda,\Gamma) \equiv N\sigma(\lambda,\Gamma) = \frac{9\pi^2}{\lambda^4N}\left(\frac{\varepsilon_r-1}{\varepsilon_r+2}\right)^2\sin^2\Gamma. \tag{4.20}$$

The units are m^{-1} sterad^{-1}.

When we assume now that the incident radiation is unpolarized (natural light), then the wave field can be regarded as composed of two linearly polarized fields perpendicular to each other, one parallel to the x- and the other parallel to the y-axis of a Cartesian coordinate system in which z is the direction of propagation. The intensities carried by each of these two components are half the total intensity I_0 of the incident radiation, or $\langle E_{0x}^2\rangle = \langle E_{0y}^2\rangle = \langle E_0^2\rangle/2$. With the definition of Γ as the angle between the direction of polarization and the direction of observation, applied to the two orthogonal polarizations we now get the components Γ_x and Γ_y and hence Eq. (4.16) changes simply to

$$I(\Gamma) = \frac{9c\varepsilon_0\pi^2}{4R^2\lambda^4N^2}\left(\frac{\varepsilon_r-1}{\varepsilon_r+2}\right)^2 E_0^2(\sin^2\Gamma_x + \sin^2\Gamma_y) \ . \tag{4.21}$$

The angle ϑ between the directions of propagation of the incident and the scattered waves respectively can be shown to be related to Γ_x and Γ_y by

$$\sin^2\Gamma_x + \sin^2\Gamma_y = 1 + \cos^2\vartheta \ . \tag{4.22}$$

Substituting Eq. (4.22) into (4.21), and this into the first part of (4.18), the Rayleigh scattering function for unpolarized radiation in terms of ϑ becomes

$$\gamma(\lambda, \vartheta) = \frac{9\pi^2}{2\lambda^4N}\left(\frac{\varepsilon_r-1}{\varepsilon_r+2}\right)^2(1+\cos^2\vartheta) \ . \tag{4.23}$$

By integrating over all scattering angles the total volume scattering coefficient due to Rayleigh scattering is obtained

$$\kappa_s \equiv 2\pi\int_0^\pi \gamma\sin\vartheta \, d\vartheta = \frac{24\pi^3}{\lambda^4N}\left(\frac{\varepsilon_r-1}{\varepsilon_r+2}\right)^2 \ . \tag{4.24}$$

The dipole density N in the denominators of Eq. (4.16) to (4.24) is no contradiction to intuition if Eq. (3.40) is considered. Obviously the relation

$$\kappa_s = N\sigma_s \tag{4.25}$$

is the general link between Eqs. (4.24) and (4.19). Here we tacitly assume that the scattering cross-sections of all particles are the same and only single scattering is considered (tenuous medium). Effects of multiple scattering are treated by e.g. Ishimaru (1978) and Tsang et al. (1985). The unit of κ_s is inverse metre, because κ_s represents the loss of radiation intensity due to scattering by particles while propagating a distance of one metre.

It is important to note that for many atmospheric scattering problems the effective dielectric constant is very close to unity, i.e. $|\varepsilon_r-1| \ll 1$, so that the following approximation:

$$\left(\frac{\varepsilon_r-1}{\varepsilon_r+2}\right)^2 \approx \frac{(\varepsilon_r-1)^2}{9} \approx \frac{4(n-1)^2}{9} \tag{4.26}$$

is allowed in all equations of this section when considering atmospheric scattering.

The Rayleigh scattering function γ can be expressed by the total volume scattering coefficient κ_s and the so-called phase function $p(\vartheta)$ according to

$$\gamma(\lambda, \vartheta) = \kappa_s\frac{3}{16\pi}(1+\cos^2\vartheta) \equiv \frac{\kappa_s}{4\pi}p(\vartheta) \ . \tag{4.27}$$

This definition yields

$$p(\vartheta) = \tfrac{3}{4}(1+\cos^2\vartheta) \tag{4.28}$$

as the Rayleigh phase function normalized to unity for the case when absorption can be neglected.

In this section we are looking at the scattering of radiation by particles of a very large size compared with a molecule. Therefore the macroscopic concepts of the dielectric constant and index of refraction have been introduced. When in a different derivation of the Rayleigh scattering the macroscopic concept is directly introduced (Ishimaru 1978, Chap. 2) instead of the molecular polarization treated in this text, then the physical volume, V, of the scattering particle enters consideration instead of the inverse of the number density of the dipoles. The directional and total scattering cross-section of a single particle are formulated in this concept as

$$\sigma(\Gamma) = \frac{9 \pi^2}{\lambda^4} V^2 \left(\frac{\varepsilon_r - 1}{\varepsilon_r + 2} \right)^2 \sin^2 \Gamma \tag{4.29}$$

and

$$\sigma_s = \frac{24 \pi^3}{\lambda^4} V^2 \left(\frac{\varepsilon_r - 1}{\varepsilon_r + 2} \right)^2 , \tag{4.30}$$

respectively instead of Eqs. (4.18) and (4.19). Introducing the free space wave number $k_0 = 2 \pi / \lambda$ and assuming spherical particles (with radius a and volume $V = 4 \pi a^3 / 3$), by which harmonically oscillating wave fields are scattered, we arrive at

$$\sigma(\Gamma) = k_0^4 a^6 \left(\frac{\varepsilon_r - 1}{\varepsilon_r + 2} \right)^2 \sin^2 \Gamma \tag{4.31}$$

and

$$\frac{\sigma_s}{\pi a^2} = (k_0 a)^4 \frac{8}{3} \left(\frac{\varepsilon_r - 1}{\varepsilon_r + 2} \right)^2 , \tag{4.32}$$

the latter being normalized to the geometric cross-section. So far, loss-free dielectrics have been assumed as scattering objects and when writing the permittivity, the prime for the real part has been tacitly supressed.

The same macroscopic field concept which led to the relations between scattering cross-sections and geometric dimensions of the scatterer can be applied to establish an absorption cross-section of a particle related to dielectric losses (imaginary part of the susceptibility, ε_r''), wavelength and geometric dimensions. The absorption cross-section σ_a for a dielectric body of volume V is given by the volume integral of the electromagnetic power loss inside the particle divided by the intensity of the incident radiation $I = \frac{1}{2} \varepsilon_0 c \, |E|^2$ [see Eq. (1.39)],

$$\sigma_a = \frac{\omega \varepsilon_0}{2I} \int_V \varepsilon_r'' |E_i|^2 dV , \tag{4.33}$$

where E_i is the electric field inside the medium. When we assume a spherical particle of small size ($a \ll \lambda$, Rayleigh scattering), the relation between the applied electric field E of the incident wave and the electric field E_i inside the dielectric is approximately the same as in electrostatics. The field inside is uniform and parallel to the mean direction of the applied field. Considerations comparable to those in Chapter 3.3, which led to the Mosotti field acting on a dipole within a dielectric, lead us now (Stratton 1941, Chap. 3) to

$$E_i = \frac{3}{\varepsilon_r' + 2} E \ , \qquad (4.34)$$

if the medium surrounding the dielectric sphere is assumed to be vacuum or air. The link of this relation to the Mosotti field (3.39) is also given by Stratton.

Inserting Eq. (4.34) into the integrand of (4.33) yields

$$\frac{\sigma_a}{\pi a^2} = 3 k_0 a \varepsilon_r'' \left(\frac{2}{\varepsilon_r' + 2} \right)^2 \qquad (4.35)$$

as the absorption cross-section of a lossy dielectric sphere small enough to permit the electrostatic approximation of the internal field. Again the free space wave number $k_0 = 2\pi/\lambda$ has to be taken in Eq. (4.35).

The sum of the scattering and the absorption cross-sections is called the total or the extinction cross-section

$$\sigma_E = \sigma_s + \sigma_a . \qquad (4.36)$$

The total cross-section represents the total power loss from the incident wave due to scattering and absorption by the particle.

With this we can define the albedo of a single particle. It is given by the ratio of the scattering cross-section to the total cross-section

$$A = \frac{\sigma_s}{\sigma_E} \ . \qquad (4.37)$$

Up to now we have assumed that the scattering particles are so small (radius a)

$$2\pi a \ll \lambda \ , \qquad (4.38)$$

that the scattering is according to Rayleigh. Equation (4.38) is called the Rayleigh criterion. There are cases where the relative dielectric constant of a scatterer is close to unity. In this case the field inside the scattering particle may be approximated by the incident field and consequently the expression (4.33) is reduced to

$$\sigma_a = k_0 \int_V \varepsilon_r'' dV \ , \qquad (4.39)$$

from which, in the case of a homogeneous sphere, the normalized absorption cross-section is derived as

$$\frac{\sigma_a}{\pi a^2} = \frac{4}{3} k_0 a \varepsilon_r'' \ . \tag{4.40}$$

This approximation is valid if

$$(\varepsilon_r' - 1) 2 \pi a \ll \lambda \ , \tag{4.41}$$

which is the Born criterion. In this case $2\pi a/\lambda$ may be large if $\varepsilon_r' - 1$ is sufficiently small.

Regarding Fig. 4.5, it is obvious that in the visible part of the spectrum there are only a limited number of particle types (except molecules) occurring in the atmosphere for which either the Rayleigh or the Born criterion is fulfilled, and even for microwaves these criteria are not always fulfilled if precipitation is involved.

The exact solution of the scattering of a plane wave by an isotropic homogeneous sphere was obtained by Mie. Expressed in a strongly simplified way, the scattering by a particle, the size of which is comparable to or larger than the wavelength, can be envisaged as an interference of partial waves which pass around and − in the case of dielectric media − through the particle. The analytic solution of the Mie scattering problem is complicated, and the interested reader is referred to textbooks (van de Hulst 1957, Born and Wolf 1964, Kerker 1969), where scattering by spherical particles is treated in considerable detail. Solution of particular problems such as the scattering by arbitrarily shaped objects (Yeh 1964) or numerical values of scattering by water droplets in haze, clouds and rain, as well as by dust particles may be found in the more specialized literature (Deirmendjian 1969).

For the simplest case, the homogeneous spherical scatterer, the solutions depend on the value of the dielectric constant and very sensitively on the ratio of the radius to the wavelength.

For the special case of a metallic sphere (ideal conductivity) Fig. 4.6 presents the normalized scattering cross-section $\sigma_s/\pi a^2$ as a function of the

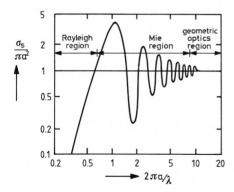

Fig. 4.6. Normalized scattering cross-section (scattering efficiency) as a function of the normalized circumference of a metal sphere. (After Siegert et al. 1963)

ratio $2\pi a/\lambda$. The dielectric constant of a conductor is imaginary and proportional to the conductivity. The index of refraction is a complex number with equal values of the real and imaginary parts $n = (1-i)\sqrt{\varepsilon_r''/2}$. From Fig. 4.6 we can distinguish three regions: For $2\pi a/\lambda < 0.5$, the curve obeys the Rayleigh approximation, i.e. $\sigma_s/\pi a^2 \propto (a/\lambda)^4$. At about $2\pi a \approx \lambda$ the maximum scatter cross-section occurs due to the Mie (or "resonant") scattering. For larger ratios of circumference to wavelength up to about 10, (again due to Mie scattering) constructive and destructive interferences of the partial waves add up vectorially, so that σ_s shows an oscillatory behaviour slowly damped out as higher modes become involved at higher a/λ values. Finally, for $2\pi a/\lambda > 10$ we reach the asymptotic property of Mie scattering, the simple geometric optics solution $\sigma_s = \pi a^2$.

If the spherical particles are from a medium with complex dielectric constant like water, the curves of scattering cross-sections versus particle size are less regular, owing to interference of surface waves travelling around the particle and the wave mode which can pass through its volume. In addition, part of the wave energy will be absorbed when passing through the particle. As an example, the computed values of the normalized cross-section for backscattering $\sigma_B/\pi a^2$ as a function of $2\pi a/\lambda$ (and as a function of the radius a for two wavelengths used in Laser backscatter experiments) are presented in Fig. 4.7.

The σ_B is the differential scattering cross-section $\sigma(\vartheta)$ for $\vartheta = \pi$. Various imaginary parts of the complex index of refraction have been assumed, the real part ($n' = 1.5$) being in the range between pure water and pure dust particles which has been determined as the average index of refraction of tropospheric non-urban haze and dust (Collis and Russell 1976b). The asymptotic value of $\sigma_B/\pi a^2$ of a sphere is given, from the geometric point of view, by $(4\pi)^{-1}$ and, for a dielectric medium, [Eq. (1.51)] leads finally to

$$\frac{1}{4\pi}\left(\frac{n-1}{n+1}\right)^2 .$$

In the practical application of remote sensing, the scattering particles not only show a complex index of refraction but exhibit irregular shapes (e.g. snow flakes) and a wide range of sizes. Often even the dielectric properties are strongly dependent on the wavelength, e.g. in the case of rain droplets sensed by microwaves. The resulting volume scattering corresponds to a mixture of curves like those drawn in Fig. 4.7.

In connection with the size dependence of the scattering mechanisms it is worthwhile to review briefly the typical size distributions of a few important kinds of scattering particles in the atmosphere. For rain there are various models to relate the range of drop sizes to the rainfall rate. An empirical expression based on long-term observations of drop-size distributions at the ground for rainfall intensities between 1 and 23 mm h^{-1} has been developed by Marshall and Palmer (1948). The number of drops of radius a, per unit volume, per drop-diameter interval is given by

$$N_r(p, a) = N_{0r}\exp(-Xa) . \tag{4.42}$$

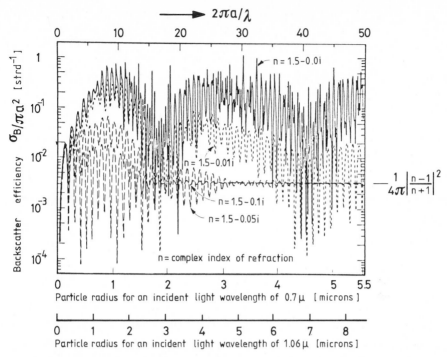

Fig. 4.7. Dependence of the Mie backscattering efficiency as a function of particle radius for two given laser wavelengths and as a function of the normalized circumference. Different imaginary parts of the complex index of refraction have been assumed in addition to the real part $n' = 1.5$, a typical value for haze and dust. (Collis and Russell 1976 b)

Here $N_{0r} = 8 \times 10^6 \, \mathrm{m}^{-4}$, $X = 8200 \, p^{-0.21} \, \mathrm{m}^{-1}$, p is the rainfall rate in mm h^{-1} and a is the drop radius. This relation is adequate as a long-term average, but for an individual shower it usually overestimates the observed values for the smallest and for the largest radii. Laws and Parsons (1943) have developed a relation which describes the size distributions of individual showers more reliably (Fig. 4.8).

One of the most accurate descriptions of "instant" rain drop distributions as they are relevant in remote sensing was achieved by Joss and Gori (1978). A more detailed parametrization led them to

$$N_{JG}(p, a) = N_{0JG} \exp(-X_{JG}a)[1 + C(X_0 - X_{JG}a)^2]^{-1} \, ,$$

where their parameters vary typically as follows: N_{0JG} from 10×10^6 to $15 \times 10^6 \, \mathrm{m}^{-4}$, X_0 from 5 to 6, X_{JG} from 3×10^3 to $8.2 \times 10^3 \, \mathrm{m}^{-1}$ and C from 0.13 to 0.4 for observed rain rates of 100 mm h^{-1} and 1 mm h^{-1} respectively in a thunderstorm situation. The resulting instant distributions (accumulation time 1 min or less) deviate strongly from exponential distributions (4.42) in the direction of monodispersity for thunderstorm as well as for widespread rain. The drop radius corresponding to the maximum of the distribution is given by

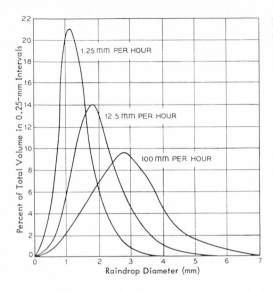

$$a_{\max} \approx X_0/X_{JG} \; ,$$

and the parameter C determines the amount of curvature (Joss and Gori 1978).

The transition between rain and fog (clouds) can be characterized by the velocity of the drops which is obviously related to their size. In Fig. 4.9 the terminal velocity as a function of the drop radius is presented after Fletcher (1962). One can roughly distinguish that the majority of cloud droplets must be in the size range below 100 μ, while rain drops (terminal velocities > 1 mg s^{-1}) must be larger than about 100 μ. Deirmandjian (1969) formulated a drop-size distribution for clouds

$$N_c(a) = N_{0c} a^\alpha \exp(-ba^\gamma) \; , \tag{4.43}$$

Table 4.3. Properties of standard cloud models. (After Ulaby et al. 1981)

Descriptive cloud name	Cloud base (m)	Height top (m)	Mass density liquid H_2O (gm^{-3})	Mode radius of distribution a_c (μm)	Shape parameters α	γ	Principal composition
Cirrostratus, mid.-lat.	5000	7000	0.10	40.0	6.0	0.5	Ice
Low-lying stratus	500	1000	0.25	10.0	6.0	1.0	Water
Fog layer	0	50	0.15	20.0	7.0	2.0	Water
Haze, heavy	0	1500	10^{-3}	0.05	1.0	0.5	Water
Fair-weather cumulus	500	1000	0.50	10.0	6.0	0.5	Water
Cumulus congestus	1600	2000	0.80	20.0	5.0	0.3	Water

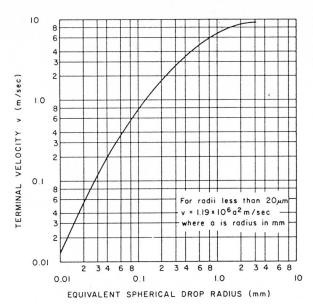

Fig. 4.9. Terminal velocity of water drops in air at 760 mm pressure and temperature of 20 °C. (Fletcher 1962)

which was called a modified gamma function since it reduces to the gamma distribution when $\gamma = 1$. For a given drop size distribution, α, b, γ and N_{0c}, are positive real constants and are related to the physical properties of the cloud. For various standard cloud models, the constants α and γ, as well as typical mean drop radii, can be read from Table 4.3. Here the most significant parameters describing the various cloud types as cloud base and height, the liquid water content and the drop size characteristics are given for standard cloud models. The constant b of Eq. (4.43) is related to α, γ and the mode radius a_c of Table 4.3 by

$$(a_c)^\gamma = \frac{\alpha}{b\gamma} , \tag{4.44}$$

with $N_c(a_c) = N_{0c}a_c^\alpha \exp(-\alpha/\gamma)$ (Deirmendjian 1969). Here a_c is the radius corresponding to the peak value of $N_c(a)$. The constant b is determined by a_c (obtainable from particle counts) provided α and γ are fixed or otherwise obtained from the shape of experimental distribution curves.

Figure 4.10 represents normalized drop-size distributions for some cloud types, the parameters of which are given in Table 4.3. One can recognize that all cloud types have a maximum number of droplets in a range of 10 to 30 μ radius. This is just the wavelength range which was earlier defined as "thermal infrared", a wavelength (≈ 10 μm) very useful for either observing the temperature and its variations at the surface of the earth or (≈ 15 μm) for the detection of carbon dioxide in the atmosphere. Mie scattering will cause a large extinction of this infrared radiation, making clouds impenetrable for these waves. One can also recognize from Fig. 4.10 that the size distributions

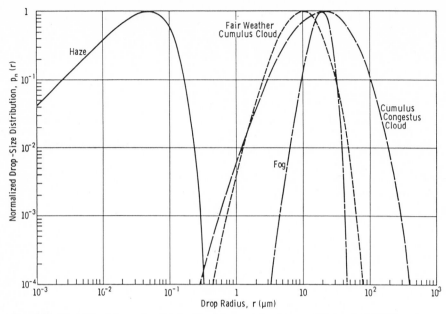

Fig. 4.10. Normalized drop size distribution for some cloud types. (Ulaby et al. 1981)

of haze particles and cloud particles are distinctly different, the maximum being shifted by two orders of magnitudes towards smaller radii.

This brings us to the size distribution of aerosols. Bolle (1982) summarized two current model distributions, one again being the modified gamma function (4.43), the other a power law distribution

$$N_P(a) = N_{0P} a^{-\beta} ,\qquad\qquad (4.45)$$

where N_{0P} depends on the concentration and β determines the slope of the distribution curve. According to Bolle, the value $\beta \approx 3$ can generally be accepted over a range of radii 0.1 µm $< a <$ 1 µm.

The effect of these size distributions on the propagation of visible light is summarized in Table 4.4.

In a clear atmosphere the resultant frequency dependence is typically between $\lambda^{-0.7}$ and $\lambda^{-2.0}$.

In order to visualize the relative efficiency of scattering, Fig. 4.11 presents the typical range of atmospheric scattering between the curves of Rayleigh and non-selective scattering. In order to avoid the scatter of light by haze in photographic applications, one has to apply haze-reduction filters with a sharp spectral cut-off, absorbing completely below about 0.4 µm. Haze-reduction films also exist which have decreasing sensitivity with decreasing wavelength and absorb completely below 0.4 µm.

The effects of the Rayleigh/Mie scattering over the wide range of wavelengths between the visible and the microwave regions caused by the wide size distributions of fogs of different density is presented in Fig. 4.12.

Table 4.4. Wavelength dependence of scattering processes for various types of particle in the atmosphere

Type of particle	Particle radius in wavelength of visible light	Dominant scatter process	Wavelength dependence of the resultant scattering
Air molecules	$\ll \lambda$	Rayleigh	λ^{-4}
Smoke Haze Aerosols	$0.1\,\lambda - 10\,\lambda$	Mie	$\lambda^{0} - \lambda^{-4}$
Dust Fog Clouds	$> 10\,\lambda$	Unselective (geometric optics)	λ^{0}

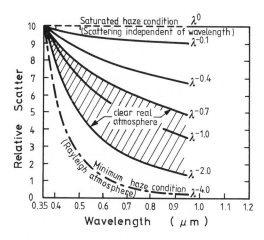

Fig. 4.11. Relative scatter as a function of wavelength for various magnitudes of atmospheric haze. (Slater et al. 1983)

The total attenuation in decibel per kilometre as a function of wavelength is given for optical visibilities ranging from only 30 m up to kilometres. Because of the absence of droplets larger than about 100 μm even in the cases of very dense fog, the attenuation above 1 mm wavelength is determined by the Rayleigh component of scattering and by absorption due to dielectric losses.

The drop-size distributions of rain are shifted to diameters of millimetres, so that millimetre and centimetre waves are subject to strong (Mie) scattering.

Figure 4.13 presents attenuation curves as a function of frequency as compiled from various sources.

As introduced earlier, the extinction of radiation is caused by two effects, the scattering and the absorption. The extinction cross-section σ_E is the sum of the scattering σ_s and the absorption σ_a cross-sections of a particle. In the simplest case of volume scattering and absorption, the attenuation per unit of

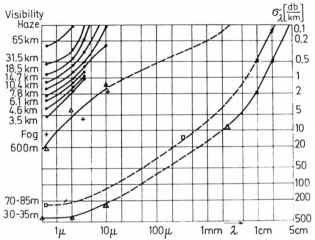

Fig. 4.12. Optical visibility and transmission loss in a horizontal path with haze or fog. (F. Kneu-buehl personal communication 1974)

distance is simply the product of the number density of scattering particles N and, if different kinds and different sizes of particle are considered, the mean extinction cross-section

$$\kappa = N\langle\sigma_E\rangle \tag{4.46}$$

in units of m^{-1} if the units of N and σ are m^{-3} and m^2 respectively. If no other effects (e.g. dispersion of the radiation beam by spatially varying index of refraction) cause attenuation to a propagating plane wave, we can identify the attenuation (4.46) with the macroscopic concept of the attenuation constant, $2k''$ for intensity, as formulated in Eq. (1.58), where k'' is the imaginary part of the complex wave number (1.54). In Figs. 4.12 and 4.13 the attenuations are presented on a logarithmic scale and the units of dB km^{-1} are used. The decibel unit for attenuation has been introduced by Eq. (2.33). The visibility, as used in Fig. 4.12, is defined by the reciprocal attenuation

$$V = \frac{1}{N\langle\sigma_E\rangle} \tag{4.47}$$

in units of metre. The albedo was introduced in (4.37) by the ratio between scattering cross-section and extinction cross-section.

Figure 4.14 shows the attenuation $N\langle\sigma_E\rangle$ and the backscattering $N\langle\sigma_B\rangle$ both in units of dB km^{-1} as well as the albedo and the half power beam width ϑ_P of the scattering pattern (proportional to the ratio of wavelength and particle size) for a typical example of a water cloud (cumulus) with a number density of $10^8\,m^{-3}$ and with $0.062\,g\,m^{-3}$ water content. The attenuation is considerable at optical wavelengths, mainly due to scattering, as the albedo is close to 1 in this range. At microwaves there is hardly any attenuation by

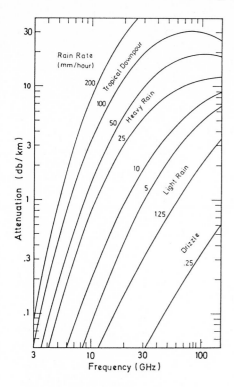

Fig. 4.13. Horizontal path attenuation at various rain rates. (Schanda 1976)

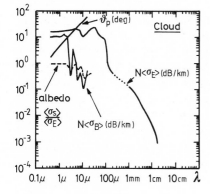

Fig. 4.14. Attenuation coefficient due to the extinction cross-section, the backscatter coefficient, the albedo and the half power beam width as a function of wavelength for a cloud. (Ishimaru 1978)

clouds, as we have already seen in Fig. 4.12 for fog. The number density N in fog and clouds may vary from $10^6\,\mathrm{m}^{-3}$ to $10^9\,\mathrm{m}^{-3}$ and the water content typically ranges from $0.03\,\mathrm{g\,m}^{-3}$ for light fog to $2\,\mathrm{g\,m}^{-3}$ for heavy fog, with corresponding visibility between 1 km and 30 m.

At this point we can refer to the scattering concept in the previous section. The albedo is large $(A \to 1)$ if $\langle \sigma_a \rangle \ll \langle \sigma_s \rangle$ a situation where elastic scattering is the dominating effect. If the albedo is small $(A \to 0)$, i.e. if $\langle \sigma_a \rangle \gg \langle \sigma_s \rangle$, the

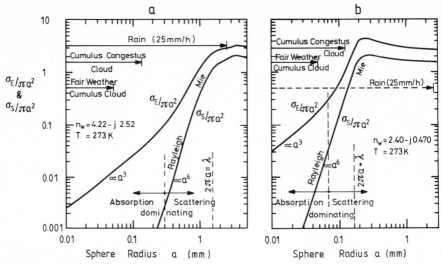

Fig. 4.15. Efficiency factors for scattering and extinction by a water sphere as a function of the drop radius at **a** 30 GHz and **b** 300 GHz. (Fraser et al. 1975)

non-elastic scattering (absorption) is the dominating effect. In general one has $0 < A < 1$ the albedo at some value between the extremes (see Table 1.2), i.e. a combination of elastic (Mie or Rayleigh-type) scattering and non-elastic scattering (absorption).

Figure 4.15a and b presents the so-called Mie-efficiency factors $\langle \sigma_s \rangle / \pi a^2$ for scattering and $\langle \sigma_E \rangle / \pi a^2$ for extinction as a function of the drop radius for two different wavelengths in the microwave region. Because of the steep decrease (proportional to volume squared) of the Rayleigh scattering with decreasing radius, the absorption (proportional to the volume) is the dominating process in the region of small droplets (clouds, fog). However, in the size range of raindrops the scattering becomes about equally as important as the absorption. The steepnesses of the Rayleigh branches in the figures are only proportional to a (absorption) and to a^4 (Rayleigh scattering) because of the normalization of volume scattering and extinction coefficients by the geometric cross-section πa^2.

4.3 Backscattering from Rough Surfaces

For the so-called active methods of remote sensing of the surface of the earth – radar and lidar – it is obvious that the scatter behaviour of rough surfaces in general, and features of the portion of radiation which is reflected back to the sensor in particular, are of fundamental interest. But also for the observation of the earth surface by radiometric methods, i.e. measuring only the natural radiation of an object in a given environment, the scatter and absorption properties of rough surfaces enter the model consideration via the albedo and the resulting emissivity in a fundamental way.

In the first chapter we treated the case of radiation reflected by, and transmitted through a plane surface. In this connection we discussed the Fresnel formulae and presented the resulting reflection coefficients for two orthogonal wave polarizations at a plane surface of a lossless medium. For the more general case of a rough boundary, for example between air and solid earth, the radiation incident on the natural surface will partly be absorbed by the medium below the surface and partly scattered, as specular reflection or diffuse dispersion into all directions. The rougher the surface, the larger the diffusely scattered portion will be. The larger the difference of material properties of air and earth — say the index of refraction of the soil — the stronger the scattered radiation will be as compared with the radiation entering the sub-surface medium [see Eqs. (1.51)]. Before introducing some models treating the scattering in a more or less approximative but quantitative way, Fig. 4.16 may serve to visualize how an incident plane coherent wave is either specularly reflected by a plane surface or more and more diffusely scattered the rougher the surface. For the slightly rough surface the angular radiation pattern consists of the following components: first, a reflected component in the specular direction with a magnitude, yet smaller than that for a plane surface, (this component is also called the "coherent" scattering component, because the phase front of a coherent wave is conserved); second, a diffusely scattered component, (also called "non-coherent" scattering component), consisting of power scattered in all directions, the phase coherence deteriorated or even destroyed, and its magnitude smaller than that of the specular component. The two limiting cases, the plane and the ideal rough (Lambertian) surfaces, yield as their respective angular patterns of the scattered radiation a delta function centred at the specular direction and a diffuse scattering uniformly distributed in the upper half space. The term "roughness" or "degree of roughness" is always understood here as to be measured in wavelengths. A given surface may appear rough at optical wavelengths and may appear smooth at microwaves.

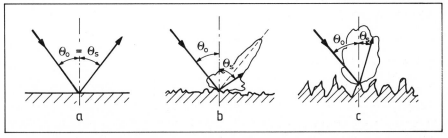

Fig. 4.16. Specular and diffuse components of the radiation scattered at **a** perfect plane, **b** slightly rough, **c** very rough surfaces

Profound reviews on the analysis of rough surface scattering are available in the literature, e.g. by Beckmann and Spizzichino (1963), a very general and fundamental approach, by Barrick (1970), primarily for microwave (radar) applications, and by Leader (1979), for scattering of laser light. The utilization of surface scattering concepts in remote sensing is much more developed in the microwave than in the optical range of the spectrum. This is partly because radar systems are all-weather day-and-night sensors which measure the backscattering of artificial radiation from the earth's surface and allow sensitive discrimination between different surface properties.

To define the geometric situation of surface scattering, Fig. 4.17 gives a schematic presentation of the relevant quantities. The intensity I_{0i} of the incident radiation with the direction i of the electric field vector (the polarization) arrives through an elementary solid angle $d\Omega_0$ at the zenith angle Θ_0 and azimuth angle φ_0. It causes a radiation power dP_{0i} at the element area dA of the surface

$$dP_{0i} = I_{0i}\, dA \cos \Theta_0 = I_{0i} R^2 d\Omega_0 \ . \tag{4.48}$$

The surface element dA is related to the distance R of the source of the incident radiation by the geometric relation

$$dA = R^2 d\Omega_0 / \cos \Theta_0 \ . \tag{4.49}$$

The power dP_{0i} at the surface may be reradiated (scattered) in all directions, an infinitesimal fraction of it, $d^2 P_{sj}$, into the elementary solid angle $d\Omega_s$ at Θ_s and φ_s, with a polarization j. This elementary fraction of the scattered power is given by

$$d^2 P_{sj} = dP_{0i} \gamma_{ji}(\Theta_0, \varphi_0; \Theta_s, \varphi_s)\, d\Omega_s / 4\pi \ . \tag{4.50}$$

The quantity γ_{ji} is the bistatic scattering coefficient relating the radiation scattered into the direction Θ_s, φ_s at polarization j with the radiation incident onto dA from direction Θ_0, φ_0 at polarization i. We are dealing with surface scattering here and in contrast to the Rayleigh scattering function (4.20), the scattering coefficient γ_{ji} is a dimensionless quantity. It will be useful to distinguish between the power scattered with no change in polarization and that with changed polarizations with respect to the polarization of the incident radiation. The scattering coefficients γ_{ii} and γ_{ji} respectively have this particular meaning.

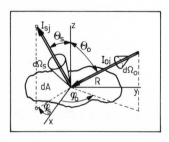

Fig. 4.17. Geometric relations between the intensities of incident I_{0i} and scattered radiation I_{sj} on a surface area dA (schematically)

All the notations within this section are obviously deduced from radar terminology (see e.g. Moore 1983). Equation (4.50) can be regarded as the differential form of the radar equation for a bistatic arrangement, i.e. where the radar transmitter and receiver are at different locations but − for simplicity of the following expression − both antennae being directed towards the same scattering object. The radar equation relates the received power P_r at the detector to the transmitted power P_t and incorporates the scattering properties of the object, the gains G_r, G_t, or the surface areas A_r, A_t of both the receiving and transmitting antennae and the radiation losses due to the distances R_r, R_t involved. It reads

$$P_r = \left[P_t \frac{G_t}{4\pi R_t^2} \right] [A_s r_s G_s] \left[\frac{A_r}{4\pi R_r^2} \right] , \qquad (4.51)$$

where the brackets group the properties of the transmitter, of the scattering object and of the receiver. The object properties are: the effective area of the object A_s, i.e. that effective cross-section of the incident wave front from which all power is removed by the object; the fraction of the radiation intensity incident on A_s which is reradiated, i.e. not absorbed, r_s; the "gain" G_s of the scatterer as determined in the direction of the receiver. The factors $(4\pi R_t^2)^{-1}$ and $(4\pi R_r^2)^{-1}$ represent the free space losses due to the distances R_t from the transmitter to the object and R_r from the scatterer to the receiver. The gain G of an antenna of given surface area A is dependent on the wavelength and is given by

$$G = 4\pi\eta A/\lambda^2 , \qquad (4.52)$$

where η is an efficiency factor, usually ranging between 0.5 and 1. For the sake of simplicity, we shall take $\eta = 1$ in the following. The properties of the scattering object embraced in Eq. (4.51) can be combined to

$$\sigma = A_s r_s G_s , \qquad (4.53)$$

where σ is the radar scattering cross-section. In the following we shall use the concept of reflection coefficient r only in connection with intensity. Therefore we will delete the subscript I, this in contrast to definition (1.51).

In monostatic radar the transmitting and receiving antennae are co-located and a single antenna is used for transmitting and receiving ($R_t = R_r$ and $A_t = A_r$). Using Eqs. (4.52) and (4.53), the radar equation for the monostatic case follows immediately from Eq. (4.51)

$$P_r = P_t \frac{A_r^2 \sigma}{4\pi\lambda^2 R_r^4} . \qquad (4.54)$$

The dependence of the received power on the fourth power of range is a severe limitation of radar for satellite-borne remote sensing applications, which cannot always be overcome by increasing the antenna size (the diameter of a circularly assumed antenna enters also with the fourth power) or by decreasing the wavelength. The scattering cross-section is of course a wavelength dependent factor in (4.54). In remote sensing, most of the targets to be investigated by radar are distributed objects such as the sea surface or agricultural fields, often larger than can be illuminated by the radar beam. Most of those surfaces can be regarded as being composed of a collection of statistically identical objects of differential size (surface area dA) each with radar cross-section $d\sigma$. The differential form of (4.54),

$$dP_r = P_t \frac{A_r^2}{4\pi\lambda^2 R_r^4} d\sigma \; , \tag{4.55}$$

is only meaningful if dP_r and $d\sigma$ are average quantities taken over all the differential objects in question, because they are identical only in the statistical sense. When comparing (4.55) with (4.50) one has to bear in mind that – essentially – we changed from dP_{0i} to $P_t dA$ (via σ) and we replaced the differential solid angle $d\Omega_s$ by the geometrically determined A_r/R_r^2. We can rewrite $d\sigma$ by $\sigma_{ji}^0 dA$ with σ_{ji}^0 the average back-scattering cross-section per unit surface area (dimension-less), where again the subscripts j and i respectively are used to characterize the polarizations of the received and transmitted radiation. Thus we can rewrite Eq. (4.55) in a form widely used in the radar remote sensing community

$$dP_r = P_t \frac{A_r^2 \sigma_{ji}^0}{4\pi\lambda^2 R_r^4} dA \; . \tag{4.56}$$

When the incidence angle of the radar beam on a given surface is changed, the illuminated surface area is also changed. Therefore it is more convenient to relate the characteristics of the average surface scattering to the differential beam cross-section $dB = \cos\Theta_0 \, dA$ instead of the surface cross-section dA, Θ_0 being the angle between the surface normal and the radar beam. Hence we shall use γ_{ji}^0 (dimension-less or expressed in decibel) instead of $\sigma_{ji}^0 dA$. With this we are able to relate the previously introduced bistatic scattering coefficient γ_{ji} with the scattering cross-section per unit surface σ_{ji}^0 for the special case of backscattering (superscript ^{0})

$$\gamma_{ji}^0 = \frac{\sigma_{ji}^0}{\cos\Theta_0} \; . \tag{4.57}$$

This backscatter coefficient γ_{ji}^0, as special case of the scattering coefficient, can be defined in terms of incidence and scattering angle as follows:

$$\gamma_{ji}^0 \equiv \gamma_{ji}(\Theta_0, \varphi_0; \Theta_0, \varphi_0 \pm \pi) \; .$$

The radar concept of treating the scattering processes in remote sensing applications is elaborated in great detail by Ulaby et al. (1981, 1982) and by Tsang et al. (1985).

Now we consider the monostatic case (backscattering only) to connect this concept of the radar equation to the idea of differential scattering equation (4.50), the latter being illustrated by Fig. 4.17. This connection can easily be established by equating (4.50) and (4.56) in pieces. Making use of Eq. (4.52) we find

$$I_{0i} = \frac{P_t}{4\pi R_r^2} \frac{4\pi A_r}{\lambda^2} \, , \tag{4.58}$$

when grouped in factors of isotropically diluted transmitter power at distance R_r and the gain of the antenna. From geometrical arguments it follows that

$$d\Omega_s = A_s/R_r^2 \, . \tag{4.59}$$

The rest is fulfilled by Eq. (4.57).

We proceed now from Eq. (4.50) with the general bistatic scattering. The total power scattered by the surface element dA into all directions can be found by integrating d^2P_s over the upper half-space; both polarizations of the scattered radiation, the conserved and the depolarized components have to be considered. Obviously this integral has to be identical to the product of incident power dP_{0i} and reflectivity $r_i(\Theta_0, \varphi_0)$ for polarization i and incidence angle Θ_0 and φ_0

$$dP_s = r_i(\Theta_0, \varphi_0)\,dP_{0i} = \int dP_{0i}(\gamma_{ii} + \gamma_{ji})\,d\Omega_s/4\pi \, . \tag{4.60}$$

Of course, dP_s contains contributions of both orthogonal polarizations, the conserved polarization due to $\gamma_{ii}\,dP_{0i}$ and the depolarized due to $\gamma_{ji}\,dP_{0i}$. Therefore, r_i represents the reflection of intensity at the like (i) and cross (j) polarization if the incident intensity is assumed to be polarized along the i-direction. Independence of the incident power on the scattering angles Θ_s, φ_s allows us to write

$$r_i(\Theta_0, \varphi_0) = \frac{1}{4\pi} \int [\gamma_{ii}(\Theta_0, \varphi_0; \Theta_s, \varphi_s) + \gamma_{ji}(\Theta_0, \varphi_0; \Theta_s, \varphi_s)]\,d\Omega_s \, . \tag{4.61}$$

The integral is to be performed over the upper half space (solid angle 2π). If we regard now the reversed observing situation, that is unpolarized radiation incident from any directions and the scattered radiation observed in one direction Θ_0, φ_0 and in only one specific polarization (i), we discover that Eq. (4.61) is identical to the albedo of the surface. For an ideally scattering surface (a substantial jump of the permittivity or a high conductivity at the interface) (4.61) will yield $r_i \approx 1$. Natural surfaces almost never have such high values of the albedo, either in the visible (see Table 1.2) or in the microwave or any other spectral range applied in remote sensing. As a consequence we may conclude that

$$a_i(\Theta_0, \varphi_0) = 1 - r_i(\Theta_0, \varphi_0) \tag{4.62}$$

represents the portion of the incident radiative power which is absorbed by the medium below the surface in question.

The integrand in Eq. (4.61) is determined by the surface roughness and the sub-surface medium and so are, consequently, the reflectivity r_i and the absorptivity a_i. Two extreme cases are specular reflection at a plane surface and scattering at a Lambertian, as a case of a perfectly rough surface. These two cases are characterized by

$$(\gamma_{ii} + \gamma_{ji})_{\text{specular}} = \frac{4\pi}{\sin\Theta_0} r_{Ii}(\Theta_0)\, \delta(\Theta - \Theta_0)\, \delta(\varphi - \varphi_0)$$

$$(\gamma_{ii} + \gamma_{ji})_{\text{Lambert}} = \gamma_0 \cos\Theta_0 \; ,$$

(4.63)

where r_{Ii} is given by Eq. (1.52) and δ-functions determine the angular dependence of the specular case; γ_0 is the scattering coefficient for $\Theta_0 = 0$ in the Lambertian case. Both functions alternately inserted in Eq. (4.61) and integrated over the upper half space yield

$$r_{i\,\text{specular}}(\Theta_0, \varphi_0) = r_{Ii}(\Theta_0)$$

$$r_{i\,\text{Lambert}}(\Theta_0, \varphi_0) = \gamma_0/4 \; ,$$

(4.64)

respectively.

Before discussing more current models of scattering at rough surfaces, we have to define a criterion for smooth versus rough. With the geometry of Fig. 4.18 we can compute the phase difference $\Delta\phi$ of two rays reflected at two different levels one standard height deviation h apart; $\Delta\phi$ is given by

$$\Delta\phi = 2h\frac{2\pi}{\lambda}\cos\Theta_0 \; .$$

(4.65)

The Rayleigh criterion states that for a surface to be smooth this phase difference has to fulfil the condition $\Delta\phi < \pi/2$, which yields

$$h < \frac{\lambda}{8\cos\Theta_0} \; .$$

(4.66)

The Rayleigh criterion is useful as a first-order classifier of surface roughness or smoothness, but for modelling the scattering behaviour of natural surfaces

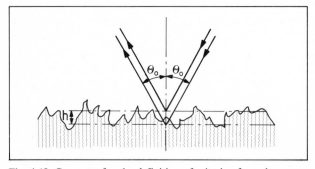

Fig. 4.18. Geometry for the definition of criteria of roughness

where the wavelength in the most important cases is of the order of h, a more stringent criterion is needed. Starting from the practice of microwave antenna techniques, Ulaby et al. (1982) propose the Fraunhofer criterion, which states that the validity of the far field (Fraunhofer) approximation in diffraction starts at distances from the diffracting aperture (antenna) where the maximum phase difference between two rays is $\Delta\phi < \pi/8$. Thus the Fraunhofer criterion for a surface to be smooth is

$$h < \frac{\lambda}{32 \cos \Theta_0} . \tag{4.67}$$

These definitions are important for the two models of rough surface scattering to be discussed in the following.

As the first example of scattering models of rough surfaces let us consider the concept of scattering from facets. This facet or tangent-plane model has already been introduced by Beckmann and Spizzichino (1963). One can approximate certain types of rough surface (e.g. to some extent: ocean waves) by a series of small planar facets, each tangential to the actual surface (Fig. 4.19) and fulfilling the conditions that their characteristic dimensions l_f are much larger than the wavelength λ, and that the deviations Δ_f of the facets from the real surface are much smaller than the wavelength.

The concept of scattering from facets is really more like a specular reflection than a scattering model. The radiation pattern of an individual facet depends on its size, as already discussed in connection with antennas [see Eq. (1.36) and Fig. 1.10]. The wider the facet, the narrower the main beam of the radiation scattered into the specular direction. Each surface element (facet) is a specular reflector, therefore the only way impinging radiation can be scattered back to the source (situation of a monostatic radar) is for the ray to strike the facet at normal incidence or, because of the finite width $\Delta\vartheta_f$ of the radiation pattern of a finite-sized facet, at an incidence within this beam width close to normal. In the notation of Fig. 4.19 this means $\Theta' \approx 0$. These facts can quantitatively be expressed by a probability distribution function $p(\alpha, \varphi)$ for the slope angle α and the azimuth angle φ of the facets. $p(\alpha, \varphi) \cdot d\Omega$ is then the probability of the vector normal on a facet to be oriented within a differential solid angle $d\Omega$ around the direction (α, φ), where

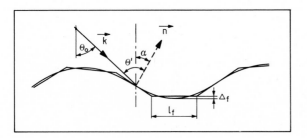

Fig. 4.19. Geometry of the facet model of a rough surface

$\langle \alpha \rangle = 0$ is the mean slope angle of the surface on large scale. Normalizing of the probability distribution requires

$$\int_0^{2\pi} \int_0^{\pi/2} p(\alpha, \varphi) \sin \alpha \, d\alpha \, d\varphi = 1 \ . \tag{4.68}$$

For a distribution of slope angles isotropic within the cut-off values $\pm \alpha_c$ and equal zero outside of $|\alpha_c|$, and an isotropic distribution of φ between 0 and 2π, the evaluation of Eq. (4.68) yields a constant probability

$$p_1(\alpha) = \begin{cases} \dfrac{1}{2\pi(1 - \cos \alpha_c)} & \text{within } \alpha = \pm \alpha_c \\[2mm] 0 & \text{elsewhere} \ . \end{cases} \tag{4.69}$$

However, on natural surfaces a probability function with a maximum around zero slopes is more likely to occur. Assuming isotropy only in azimuth φ, it can readily be proven that the function

$$p_2(\alpha) = \begin{cases} \dfrac{\cos \alpha}{\pi \sin^2 \alpha_c} & \text{within } \pm \alpha_c \\[2mm] 0 & \text{elsewhere} \end{cases} \tag{4.70}$$

fulfils the condition (4.68). In order to be applicable to the general monostatic radar configuration, α_c has to fulfil simultaneously the conditions of

normal incidence of waves on the facets $\qquad \alpha_c \geqslant \Theta_0$

and of avoiding shadow effects $\qquad \alpha_c \leqslant \dfrac{\pi}{2} - \Theta_0 \qquad (4.71)$

which result in $\qquad \alpha_c \leqslant \pi/4 \ .$

The relation between the solid angles subtended by the facet normal for a given range $d\alpha$ around the orientation towards the radar and those subtended by the propagation directions of the backscattered ray respectively can be derived from Fig. 4.20.

The probability for reflection within the solid angle $d\Omega_s$ around the direction $\alpha = \Theta_0$ and $\varphi = \pi$, for small values of $d\Omega_s$, demanded by the small $\Delta\vartheta_f$, is approximately given by

$$p(\alpha, \varphi) d\Omega_n \approx p(\Theta_0, \pi) \frac{d\Omega_s}{4} \ . \tag{4.72}$$

The reflection coefficient for radiation intensities with perpendicular incidence can be taken from Eq. (1.51)

$$r_\perp = \left(\frac{n-1}{n+1} \right)^2 \ ,$$

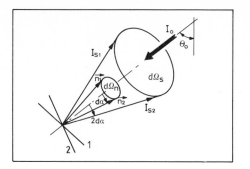

Fig. 4.20. Relation between the solid angle $d\Omega_s$ into which a wave from Θ_0 and φ_0 is specularly reflected when any facet angle within $d\Omega_n$ is admitted

when we assume the same index of refraction n below all facets and unity above. We can now use this concept in the formula for the differential backscattered power (4.50) and we have to keep in mind that in a first approximation no depolarization of the wave field occurs by the nearly perpendicular reflection at plane surfaces. Therefore we arrive at

$$d^2P_{si} = dP_{0i}\gamma_{ii}(\Theta_0, \varphi_0; \Theta_0, \varphi_0 \pm \pi) \frac{d\Omega_s}{4\pi} = dP_{0i}\, r_\perp\, p(\Theta_0, \pi) \frac{d\Omega_s}{4} \,, \quad (4.73)$$

where γ_{ii} indicates the conservation of the state of polarization and in practical cases one tends to use either horizontal or vertical polarization. From Eq. (4.73) we conclude

$$\gamma_{ii}^0(= \gamma_{hh}^0 \quad \text{or} \quad = \gamma_{vv}^0) = \pi r_\perp\, p(\Theta_0, \pi) \,. \quad (4.74)$$

The fact of negligible depolarization can be expressed by $\gamma_{ji}^0(= \gamma_{hv}^0 \text{ or } = \gamma_{vh}^0)$ $= 0$.

For the probability distribution of the slope angles Eq. (4.70) and isotropic azimuth distribution Eq. (4.74) becomes

$$\gamma_{ii}^0 = r_\perp \frac{\cos \Theta_0}{\sin^2 \alpha_c} \,. \quad (4.75)$$

We can extend for a moment Eq. (4.70) to $\alpha_c \to \pi/2$, which yields

$$\gamma_{ii}^{0L} = r_\perp \cos \Theta_0 \,, \quad (4.76)$$

a backscatter coefficient of the "Lambertian" surface [relation (4.57) leads to $\sigma_{ii}^{0L} = r_\perp \cos^2 \Theta_0$]. Equation (4.76) compares favourably with the more general formulation (4.63). With this formula, however, we have disobeyed the inherent limitation of our approach (4.71) the necessity to avoid shadowing, in other words Eq. (4.76) represents a Lambertian surface for which only viewing close to the vertical ($\Theta_0 \approx 0$) is allowed. If consequently $\Theta_0 \approx 0$ is inserted, Eq. (4.76) yields $\gamma_{ii}^{0L} \approx r_\perp$. This result seems, at a first glance, to be an overestimation of γ_{ii}^0, because any rough surface is thought to scatter radiation away from the main beam direction. However, on a rough surface, there is also radiation scattered back to the sensor from directions which would not contribute on a specularly reflecting plane surface. For the more realistic range of slope angles, ($\alpha_c \leqslant \pi/4$), one can easily prove that inserting Eq. (4.75) into Eq. (4.73) and integrating over Ω_s (the limit of Θ_0, being $2\alpha_c$) yields

$$dP_{si} = r_\perp dP_{0i} \ ,$$

the same result as for the limiting case of a plane-reflecting surface. When we step even further and apply the probability function of the isotropic slope distribution (4.69) for $\alpha_c \to \dfrac{\pi}{2}$ we arrive at

$$\gamma_{ii}^{0P} = \frac{r_\perp}{2}$$

independent of Θ_0. This (for the perfect rough surface in terms of back-scattering) looks not unrealistic when compared with Eq. (4.76) because the slope distribution (4.69) causes more radiation to be scattered away from the direction Θ_0 than distribution (4.70), as long as, in both cases, the incidence angle Θ_0 is kept close to nadir. It must be admitted that these simple facet models lack a reliable experimental verification. The models are incomplete in particular with reference to the shadow effect, (Brown 1984), and to multiple reflections at steep α_c values, neither of which are considered here. Nevertheless, the approximate agreement is encouraging at least for the application of Eq. (4.75) in a first-order estimation of the scattering behaviour of certain simple classes of roughness. A detailed and thorough treatment of models for scattering of radar waves at almost any kind of surface has been given by Ulaby et al. (1982, in particular Chaps. 11 and 12).

After considering this simplified facet model based on geometric optics consideration, sometimes called the Kirchhoff method, we shall regard in the following a statistical surface roughness model (Barrick 1970).

For defining the quantities involved we use Fig. 4.21.

We assume an irregular surface height profile deviating from a horizontal plane in a random way with the standard deviation

$$h = [\langle (z(x, y) - \langle z \rangle)^2 \rangle]^{1/2} \ , \tag{4.77}$$

with x, y the coordinates in the reference plane and z the local deviation from this plane.

We simplify the considerations if we assume a weak roughness. The conditions for this are

small roughness height: $2\pi h/\lambda < 1$

gentle slopes: $\left| \dfrac{\partial z}{\partial x} \right| < 1 \ , \quad \left| \dfrac{\partial z}{\partial y} \right| < 1 \tag{4.78}$

and azimuthal isotropy: $\left\langle \left(\dfrac{\partial z}{\partial x} \right)^2 \right\rangle = \left\langle \left(\dfrac{\partial z}{\partial y} \right)^2 \right\rangle .$

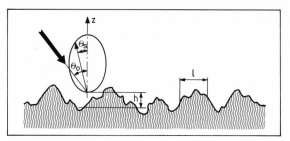

Fig. 4.21. Pattern of the radiation scattered on a rough surface with Gaussian distributed height deviations having standard height deviation h and correlation length l

For all realistic statistical models there will always be a correlation between the deviation from the reference plane $z(x, y)$ at x and y and the deviation $z(x', y')$ of a neighbouring point at coordinates x' and y'. This fact can be expressed by the correlation function normalized on the standard deviation

$$c(\rho) = \langle z(x, y) z(x', y') \rangle / h^2$$

$$\text{with} \quad \rho^2 = (x - x')^2 + (y - y')^2 \ . \tag{4.79}$$

The correlation function for a statistical variable as our rough surface deviation z is usually a monotonically decreasing function and the decrease is characterized by the correlation length l. This correlation length l along x and y of the variable z can be assumed according to various monotonically decreasing functions. Let us take a Gaussian shaped correlation function

$$c(\rho) = \exp(-\rho^2 / l^2) \ . \tag{4.80}$$

For this type of roughness, Barrick (1970) gives for the average incoherent scattering cross-section per unit surface area

$$\sigma_{ji} = \frac{4}{\pi} k_0^4 h^2 \cos^2 \Theta_0 \cos^2 \Theta_s |\alpha_{ji}|^2 \smallint \tag{4.81}$$

for the bistatic case, where $k_0 = 2\pi/\lambda$, and α_{ji} are proportional to the scattering matrix elements for all polarization states and depend on permittivity of the subsurface material, the incidence and the scattering angles. Finally the integral $\smallint$ is given by

$$\smallint \equiv 2\pi \int\limits_0^\infty \rho c(\rho) J_0(k_0 \sqrt{\xi_x^2 + \xi_y^2}\,\rho)\, d\rho \ ,$$

with $\xi_x = \sin \Theta_0 - \sin \Theta_s \cos \varphi_s \, (= 2 \cdot \sin \Theta_0$ for backscatter)
$\xi_y = \sin \Theta_0 \sin \varphi_s \, (= 0$ for backscatter due to $\varphi_s = \pi)$.

When the correlation function (4.80) is used and Eq. (4.81) is applied to the monostatic case in our notation (4.57), the backscatter coefficients for vertical and horizontal polarizations are after Barrick (1970)

$$\gamma_{vv}^0 = 4 k_0^4 h^2 l^2 \cos^3 \Theta_0 \left| \frac{(\varepsilon_r - 1)[(\varepsilon_r - 1)\sin^2\Theta_0 + \varepsilon_r]}{[\varepsilon_r \cos\Theta_0 + \sqrt{\varepsilon_r - \sin^2\Theta_0}]^2} \right|^2 \exp[-k_0^2 l^2 \sin^2\Theta_0]$$

$$\gamma_{hh}^0 = 4 k_0^4 h^2 l^2 \cos^3 \Theta_0 \left| \frac{\varepsilon_r - 1}{[\cos\Theta_0 + \sqrt{\varepsilon_r - \sin^2\Theta_0}]^2} \right|^2 \exp[-k_0^2 l^2 \sin^2\Theta_0]$$

$$\gamma_{vh}^0 = \gamma_{hv}^0 = 0 \ . \tag{4.82}$$

Computed values of $\sigma^0(\Theta_0)$, normalized by $k_0^2 h^2$ to keep the ordinate dimensionless (logarithmic scale), are presented in Figs. 4.22 and 4.23 for vertical and horizontal polarization. Curves are given for different normalized correlation lengths and various permittivities of the sub-surface medium.

The backscattering increases considerably with the permittivity increasing from the lowest values ($1 \leq \varepsilon_r \leq 5$) but it saturates rapidly for higher permittivities. Every 10 units on the ordinate is equivalent to a factor of 10 in the intensity of the backscattered signal. Small correlation lengths ($k_0 l = 0.2$) show a dependence on the incidence angle typical for very rough surfaces, comparable to what was said in connection with the "perfect" and the Lambertian scatterers. Large correlation length ($k_0 l = 5$) shows a steep dependence on Θ_0 and tends to behave as a partially flat surface. This is because the surface areas which reflect specularly are increasing and therefore the resulting radiation beamwidths are decreasing. This smoothing effect is particularly pronounced in this example with the Gaussian-shaped correlation function. The computed curves for intermediate correlation length ($k_0 l = 1$) look qualitatively very similar to those for $k_0 l = 0.2$ (same character of

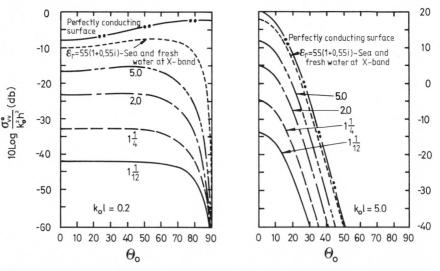

Fig. 4.22. Average incoherent backscattering cross-section per unit surface area normalized by $(2\pi h/\lambda)^2$ for slightly rough surface statistical model versus incidence angle for vertical incident and vertical scattered polarization

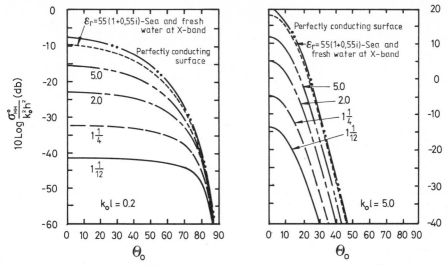

Fig. 4.23. The same as Fig. 4.22 for horizontal incident and horizontal scattered polarization. In both figures Gaussian surface height correlation with correlation length l, and $(2\pi h/\lambda)^2 < 1$ are assumed. (Barrick 1970)

angular dependence). However, the absolute values at nadir ($\Theta_0 = 0$) are on the logarithmic scale about proportionally in between the ones of $k_0 l = 0.2$ and $k_0 l = 5$.

The scattering response obtained from actual radar remote sensing measurements of the earth's surface almost never gives a model-like rough surface which can be described by a single set of roughness parameter (e.g. random height above a mean planar surface with a single mean square deviation and with a single correlation length of a well-defined correlation function additional to a single-valued permittivity of a homogeneous subsurface medium). In many applications, such as for oceanographic and agronomic purposes, the measured scattering response can be reasonably well approximated by a two-scale roughness model (i.e. one that superimposed on a large roughness a second kind of roughness at a different scale and perhaps with a different permittivity of the medium, e.g. capillary waves). Two-scale roughness is treated in some texts, such as Barrick (1970) and Ulaby et al. (1982), from whom Fig. 4.24 has been taken. Computed backscatter cross-sections per unit area, σ^0, according to a two-scale model theory, is compared with measurements at an artificially generated two-scale rough surface (see also recent studies on composite surface scattering by Fung and Chen 1984).

Considerations of this kind go far beyond the goals of these lecture notes. However, there is one more topic to which attention should be drawn.

On the natural surfaces there is always some portion of the surface predominantly flat which contributes a considerable specular reflection to the scattered radiation. This can be caused by smooth soil beneath the vegetation

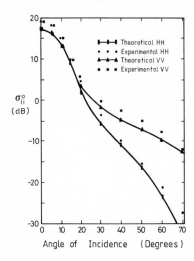

Fig. 4.24. Comparison of a theoretical two-scale roughness ("tilted-perturbed-surface") model with measured backscatter cross-section per unit area (Ulaby et al. 1982, p. 984). For this model experiment an aluminium plate was prepared for a combined large-scale ($h \approx 2.24$ mm, $l \approx 23.4$ mm) and small-scale ($h \approx 0.85$ mm, $l \approx 6.2$ mm) rough surface and measured at 25 GHz

canopy, by wide troughs between the ocean wave crests and many other reasons. The result is an angular distribution of the scattered radiation, as indicated in Fig. 4.16b and also in Fig. 4.19, consisting of a wide, diffusely scattered and a narrow, specularly reflected component. For the radar viewing angles close to nadir this latter component dominates the received radiation completely, while for viewing angles sufficiently far off nadir this specular component decreases steeply and the diffuse component remains over a wide angular region. The width of the specular regime depends on both the size of the antenna and the size of the flat pieces (large facets) at the ground because these sizes determine in an inverse dependence the widths of the respective diffraction beams. The width of the diffuse regime is essentially dependent on the particular roughness scale in height and width. An excellent experimental demonstration on how the roughness scale influences the angular dependence of the scattering cross-section has been reported by Ulaby et al. (1978) and is reproduced in Fig. 4.25.

Normalized to the wavelength of about 27 cm, the different curves are for roughness scales varying from almost flat ($h/\lambda \approx 0.04$) to moderately rough ($h/\lambda \approx 0.15$, i.e. $k_0 h \approx 1$). Below about 10° off nadir one can recognize that the flatter the surface the stronger the specular reflection, above 10° the rougher the surface the stronger the diffuse scattering.

This makes it obvious that the contributions by flat portions of the surface have to be considered. We are reminded of the Fresnel formulae (1.52) of the reflectivities for horizontally and vertically polarized waves r_{Ih} and r_{Iv} respectively, and the corresponding formula (4.63) of the scattering coefficient, both valid at the infinite plane interface between air and a dielectric. The scattering cross-section of a flat surface of finite size is derived by Barrick (1970). For rectangular flat pieces of length L_x in the plane of incidence, and of width L_y perpendicular to the plane of incidence (parallel to the horizontally polarized

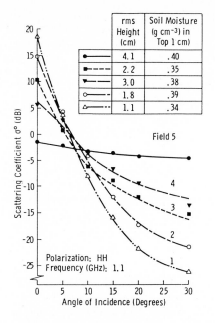

	rms Height (cm)	Soil Moisture (g cm⁻³) in Top 1 cm)
•———	4.1	.40
■----	2.2	.35
▼——	3.0	.38
○——-	1.8	.39
△——⋯	1.1	.34

Fig. 4.25. Angular pattern of σ_{hh}^0 at 1.1 GHz ($\lambda \approx 27$ cm) for five bare soil fields with different surface roughnesses. (Ulaby et al. 1978). (Copyright 1978 IEEE)

field vector), with $k_0 = 2\pi/\lambda$, and Θ_0 the incidence angle measured from nadir, the backscattering cross-section per unit surface area is given by

$$\sigma_{ii}^0 = \frac{k_0^2 L_x L_y}{\pi} \left[\frac{\sin(k_0 L_x \sin\Theta_0)}{k_0 L_x \sin\Theta_0} \right]^2 |r_{ii}(\Theta_0)\cos\Theta_0|^2 . \qquad (4.83)$$

The pair of subscripts ii stand either for hh or vv. In either of these two cases the reflectivity (Fresnel reflection coefficient) $r_{ii}(\Theta_0)$ has to be taken for horizontal and vertical polarization accordingly. The cross-polarized term σ_{ji}^0 vanishes in the backscattering situation. σ_{ii}^0 becomes larger with two-dimensional increase of size only at nadir incidence. For growing L_x the product $L_x L_y$ already becomes compensated at small incidence angles $\Theta_0 \gtrsim \lambda/4L_x$.

Although the angular behaviour is dominated by the steep $(\sin X/X)$ function, also $\cos^2\Theta_0$ tends to decrease the off-nadir backscattering, augmented for vertical polarization and reduced for horizontal polarization due to the Θ_0 dependences of $r_{Iv}(\Theta_0)$ and $r_{Ih}(\Theta_0)$ respectively (see Fig. 1.14). For this formula an infinitely narrow antenna beam is assumed, which cannot pick up any scattered wave from the ground except at the direction Θ_0. If any additional small surface roughness is superimposed on this flat plane, the coherent nature of the specular reflection is not destroyed as long as the root mean square deviation of a random height distribution much smaller than the wavelength can be assumed. The attenuation resulting from the roughness-induced scattering can be accounted for by a factor $\exp(-4k_0^2h^2\cos^2\Theta_0)$ to Eq. (4.83).

5 Transport of Radiation

5.1 The Equation of Radiative Transfer

In the previous chapters we have discussed the interaction of electromagnetic radiation with matter, using two essentially different concepts. In one concept, the medium was globally described by a phenomenological quantity of Maxwell's macroscopic picture of the material, e.g. the permittivity ε of the sub-surface medium in the case of wave scattering at rough or plane earth-air boundaries. It was possible in these situations to analyze the integral behaviour of the wave propagation, but we had to accept a lack of insight into the physical processes. Alternatively, there existed sufficient comprehension of the detailed processes, e.g. the Rayleigh and Raman scattering of radiation by molecules, but we had not yet developed a concept to describe the macroscopic action of a volume filled with molecules, each interacting individually with the radiation.

The concept of radiative transfer is an appropriate basis to fulfill this goal. In particular it is useful to formulate the absorption, scattering and creation of natural radiation within a volume filled with particles interacting with the radiation. This is very obvious in astrophysical problems, for which this concept was first developed (Chandrasekhar 1960).

Let us assume a "cloud" of particles bound – for the sake of a simpler presentation – by two parallel planes at $z = 0$ and at $z = z_B$, and a spectral radiation intensity $I_{\nu A}$ per unit of bandwidth at frequency ν entering the cloud at $z = 0$ (Fig. 5.1).

We first ask how much of this intensity will leave the cloud at $z = z_B$, if we assume that only absorption processes are effective. The volume absorption coefficient (unit m^{-1})

$$\kappa_a(\nu) = N\langle\sigma_{a\nu}\rangle \tag{5.1}$$

will in general be as frequency-dependent as the mean absorption cross-section $\langle\sigma_{a\nu}\rangle$ of the absorbing particles is frequency-dependent. N is the number density of absorbing particles. The absorption by an infinitesimally thin (dz) layer of the intensity will be an infinitesimal change

$$(dI_\nu)_a = -\kappa_a(\nu, z)I_\nu(z)dz. \tag{5.2}$$

The absorption coefficient κ_a can vary along the transmission path z, due to varying number density as well as due to the changing character of the absorption process.

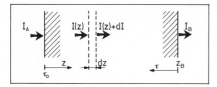

Fig. 5.1. The transfer of radiation through a slab of thickness z_B. Subscript v for the frequency dependence of the intensity omitted

The thermal agitation and other radiation processes will cause radiation to be emitted within the slab dz in amount depending on the temperature and on some other parameters if non-thermal contributions to the emission are present.

The growth rate of intensity dI_v per infinitesimal layer thickness dz can be attributed to an emission coefficient j (in units of $W\,m^{-3}\,Hz^{-1}$), as

$$(dI)_e = j(v,z)\,dz\,.\tag{5.3}$$

In sum, the change of the spectral intensity due to emission and absorption per dz is

$$\frac{dI_v}{dz} = j(v,z) - \kappa_a(v,z)I_v\,,\tag{5.4}$$

which is the equation of radiative transfer for a non-scattering medium, sometimes called Schwarzschild's equation.

If the effects of scattering cannot be neglected, there are again two terms as in Eq. (5.4), one for the loss of intensity due to radiation scattered away from the original path, and one for the gain due to radiation from other directions scattered into the direction of the observer. For the loss of intensity per element of path dz we can, according to Eq. (4.25), for tenuous distribution of scattering particles (single scattering process), deal with a total volume scattering coefficient

$$\kappa_s(v) = N\langle\sigma_{sv}\rangle\,,\tag{5.5}$$

which causes a loss rate of

$$(dI_v)_s = -\kappa_s(v,z)I_v(z)\,dz\,.\tag{5.6}$$

In the considered volume different kinds of scattering (and absorbing) particles may be present; therefore we have to use mean values, $\langle\sigma\rangle$, of the respective cross-sections.

Figure 5.2 helps to explain the relationship of Eq. (5.6) to the concepts of scattering of radiation at particles, as developed in Chapter 4.2. Usually the scattering of radiation is not isotropically distributed in all directions. We have already encountered this fact during the discussion of Rayleigh scattering and taken non-isotropy into consideration by introducing the so-called phase function (4.28) for this special case. When dI_v is taken as that part of I_v which is scattered in all directions (4π), while passing through dz [see Eq. (5.6)], then $d^2[I_v(\vartheta)]$ can be regarded as that portion of dI_v which is scattered into the direction ϑ and φ within a differential solid angle $d\Omega$ (Fig. 5.2)

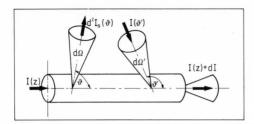

Fig. 5.2. Loss and gain of radiation by scattering

$$d^2 I_{vs}(\vartheta) = -N\langle\sigma_{sv}\rangle I_{vs} p(\cos\vartheta)\, \frac{d\Omega}{4\pi}\, dz \qquad (5.7)$$

with the phase function $p(\cos\vartheta)$ for the general case. Azimuthal symmetry (no dependence on φ) of the phase function is assumed. For purely scattering particles, the integration of Eq. (5.7) over all space angles must yield a result identical to Eq. (5.6). This condition yields the normalization of the phase function

$$\frac{1}{4\pi}\int_{4\pi} p(\cos\vartheta)\, d\Omega = \begin{cases} 1 & \text{for perfect scatterers} \\ A<1 & \text{the albedo if absorption is present} \end{cases}. \qquad (5.8)$$

The simplest examples of phase functions are:

$p(\cos\vartheta) = 1$ for isotropic scattering

$p(\cos\vartheta) = \frac{3}{4}(1+\cos^2\vartheta)$ for Rayleigh scattering

$p(\cos\vartheta) = 1 + x\cos\vartheta$ with $(-1\leqslant x\leqslant +1)$ for typical scattering by terrestrial surfaces in the visible range.

$$(5.9)$$

All examples of Eq. (5.9) are for perfect scattering. If absorption becomes involved, the right-hand sides of Eq. (5.9) have to be multiplied by the albedo A.

After this interpolation on the phase function, we are prepared to formulate the last term of the complete equation of radiation transfer, the one which takes into account the increase of radiation intensity in the direction of positive z due to scattering. All contributions by radiation from all directions (see Fig. 5.2) scattered from particles within the layer dz in such a way that the scattered radiation reenters the positive z-direction, have to be integrated over all space directions, keeping in mind the directional dependence of the scattering process according to the phase function $p(\cos\vartheta)$

$$\frac{(dI_v)_r}{dz} = \frac{N\langle\sigma_{sv}\rangle}{4\pi}\int_{4\pi} I_v(\vartheta')\, p(\cos\vartheta')\, d\Omega'. \qquad (5.10)$$

This rate of increasing spectral intensity is, for a scattering medium (e.g. atmosphere), the equivalent of the emission coefficient j of a purely absorbing medium.

The reader is reminded here of the relation between the macroscopic attenuation constant $2k''$ from Eq. (1.58) for the intensity of a plane wave and the attenuation $N(\langle\sigma_s\rangle + \langle\sigma_a\rangle)$ due to scattering and absorption by N particles per unit of volume as expressed by Eq. (4.46) and (4.36).

Let us return now to the transfer of radiation in a medium free of scattering, as expressed by Eq. (5.4). A case which is, in some sense, the opposite of a scattering medium is that of a medium in local thermodynamic equilibrium. A medium such as the earth's atmosphere is not precisely in thermodynamic equilibrium. However, in the lower atmosphere conditions prevail which are known as local thermodynamic equilibrium (LTE). This means that on local scale, molecular and atomic collision rates are much greater than radiative transition rates and the ratio of the populations of the different energy levels obeys the Boltzmann law [Eq. (2.2)], allowing the definition of a local temperature T at each point of the medium. The dimensions of this local scale have to include the volume relevant to the physical processes. These circumstances permit the emission coefficient at each point to be related to the absorption coefficient by Kirchhoff's law, so that the Planck blackbody function [Eq. (2.9)] could be employed. In the Chapter 5.2 a simple model will serve to make Kirchhoff's law plausible.

Before this step we introduce two abbreviations which will somewhat help to simplify the mathematical form of Eq. (5.4). It is general practice to call

$$S(v) = \frac{j(v)}{\kappa_a(v)} \tag{5.11}$$

the source function of incoherent radiation at frequency v and to use

$$d\tau = -\kappa_a(v,z)\,dz \tag{5.12}$$

for the (infinitesimal) optical path. This yields (see Fig. 5.1)

$$\tau = \int_0^\tau d\tau = \int_{z_B}^z \kappa_a(v,z)\,dz \tag{5.13}$$

as the optical thickness of the material between the points z and z_B also called opacity; when measured from the top of the atmosphere ($z_B \to \infty$) downward, it is called optical depth. For the quantity

$$\exp\left(-\int_z^{z_B} \kappa_a dz\right), \tag{5.14}$$

the term fractional transmission is used by the atmospheric sounding community (e.g. Houghton 1977). Applying Eq. (5.11) and (5.12), the equation of transfer (5.4) can be written as

$$\frac{dI_v}{d\tau} = I_v - S_v . \tag{5.15}$$

If thermodynamic equilibrium prevails, i.e. if a balance is established between kinetic and radiative absorption and emission processes at a given temperature, Kirchhoff's law states that the source function (the ratio of emission to absorption coefficient) is given by the Planck function for blackbody emission

$$\left(\frac{j}{\kappa_a}\right)_b = S_b(v, T) = \frac{2hv^3}{c^2}\left[\exp\left(\frac{hv}{kT}\right) - 1\right]^{-1} , \qquad (5.16)$$

i.e. the spectral radiance or radiated power in Watt per square metre per unit of solid angle and per Hertz bandwidth.

This differs from Eq. (2.9) by factors c, the speed of light, and $1/4\pi$, because there we dealt with radiation energy (Joule) per unit of volume (m^3) and per unit of bandwidth (Hz). Radiation energy per unit of volume is related to radiation power per unit of cross section and per unit solid angle by the speed of this radiation and by the unit of solid angle. In our simplified one dimensional model of radiative transfer (Fig. 5.1) we suppress the unit of solid angle, when applying Eq. (5.16).

A formal solution of Eq. (5.15) can be written down for the spectral radiation intensity arriving at the surface B

$$I_{vB} = I_{vA}e^{-\tau_0} + \int_0^{\tau_0} S_v e^{-\tau}d\tau , \qquad (5.17)$$

where use is made of Eq. (5.13) and in particular τ_0 is the optical thickness of the whole layer between $z = z_A = 0$ and $z = z_B$. If we have a homogeneous layer of material, in which $j/\kappa_a = S_v$ is independent of z (or of τ), for example an isothermal atmosphere along the path — a condition which is fulfilled only in rare exceptions — then we can execute the integration in Eq. (5.17) very easily

$$I_{vB} = I_{vA}e^{-\tau_0} + S_v(1 - e^{-\tau_0}) . \qquad (5.18)$$

The result indicates very clearly that the original radiation, entering at $z_A = 0$ into the layer of material, becomes attenuated by absorption, the transmission being $e^{-\tau_0}$. The second term on the right-hand side corresponds to the radiation emitted by the material within the layer; it accumulates to $(1 - e^{-\tau_0})$ times the source function, when reaching the surface at either z_B or $z_A = 0$.

Figure 5.3 shows these two factors which determine whether the original (or background) radiation (spectral intensity I_{vA}) is transmitted through the layer or the radiation emitted by the material itself as the source (source function S_v) dominates the observed radiation at the surface z_B. From inspection of Fig. 5.3, one can see that for $\tau_0 \leqslant 0.1$ the layer may be called transparent, while for $\tau_0 \geqslant 3$ it is opaque. A layer with

$$\tau_0 = \int_{z_B}^{z_A = 0} \kappa_a dz = 1$$

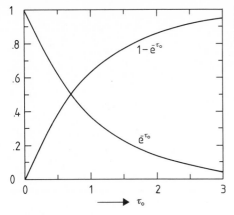

causes an attenuation of the background intensity I_A down to about 37% of its original value at $z = 0$, and the radiation emitted by the layer itself contributes about 63% of S_ν to the intensity observed at z_B.

Continuing with this simplified homogeneous and isothermal layer, we find that for a thin layer ($\tau_0 \ll 1$) there are two limiting cases for which further simplifications of (5.18) are possible.

If $I_{\nu A} \gg S_\nu$ we get $I_{\nu B} \approx I_{\nu A}(1 - \tau_0)$,

if $I_{\nu A} \ll S_\nu$ we get $I_{\nu B} \approx I_{\nu A} + \tau_0 S_\nu$.   (5.19)

In the first case only the background radiation is observed, however slightly absorbed. This absorption is characteristic of the material of the layer, e.g. a spectral line absorption; therefore this observation can be used to determine the type and (due to $\tau_0 \approx N \langle \sigma_{av} \rangle z_B$, with z_B the thickness of the layer) the quantity of intervening material causing the absorption. In the second case of Eq. (5.19), besides $I_{\nu A}$, the product of the optical thickness and the source function is measured. Because of the assumption of a homogeneous layer, this product $\tau_0 S_\nu = \kappa_a z_B \cdot j/\kappa_a = j \cdot z_B$ turns out to be the emission coefficient of the material times the thickness of the layer. If a spectral line causes the absorption, then one can determine the background spectral intensity $I_{\nu A}$ by frequency detuning the receiver. This allows the detection of the relevant quantities as $N z_B$, $j z_B$ and with S_ν also T, as long as the assumption of thermodynamic equilibrium is approximately fulfilled.

In the literature on transfer of visible and infrared radiation, it is a widespread custom to present the Planck function over a wavelength scale instead of a frequency scale, and accordingly to express it in units of power per unit of surface area per unit of wavelength per unit of solid angle. Because of Eq. (1.28) it reads as

$$S_b(\lambda, T) = 2 \frac{hc^2}{\lambda^5} \left[\exp\left(\frac{hc}{\lambda k T} \right) - 1 \right]^{-1} \qquad (5.20)$$

instead of Eq. (5.16). The different presentations cause the respective maxima to be [from (5.16)] at

$$\left.\begin{array}{l} \dfrac{\nu_{max}}{T} \approx 0.588 \times 10^{11}\, \dfrac{\text{Hz}}{\text{K}} \\[12pt] \text{or [from (5.20)] at } \lambda_{max} T \approx 2.898 \times 10^{-3}\, \text{m K} \end{array}\right\} \tag{5.21}$$

which is at about $h\nu_{max}/kT \approx 2.82$ and $hc/\lambda_{max}kT \approx 4.97$ respectively. Equation (5.21) is Wien's displacement law. At frequencies sufficiently higher than ν_{max}, e.g. above approximately 10^{14} Hz (wavelengths shorter than about 3 μm), for environmental temperatures the steep decrease of the Planck function (see Fig. 2.2) is, according to Wien's approximation

$$S_b(\nu, T) \approx 2\,\frac{h\nu^3}{c^2} \exp\left(-\frac{h\nu}{kT}\right)$$

$$\tag{5.22}$$

$$S_b(\lambda, T) \approx 2\,\frac{hc^2}{\lambda^5} \exp\left(-\frac{hc}{\lambda kT}\right)$$

in the two different presentations.

Inspection of Fig. 2.2 makes it evident that for frequencies sufficiently lower than ν_{max} the spectral radiance is proportional to the square of the frequency. One can also recognize that for a given frequency within this region $S_b(\nu, T)$ is directly proportional to the temperature. From Eq. (5.21) we may conclude that for ν (Hz) $< 10^{10}\,T(\text{K})$ the Planck function can be approximated by

$$S_b(\nu, T) \approx 2kT \left(\frac{\nu}{c}\right)^2 = 2kT/\lambda^2$$

$$\tag{5.23}$$

$$S_b(\lambda, T) \approx 2\,\frac{ckT}{\lambda^4}\,,$$

which is the Rayleigh-Jeans approximation.

Equations (5.23) are valid with satisfactory accuracy for wavelengths longer than about 100 μm, if the temperature of the radiating medium is at environmental temperature (≈ 300 K), or for wavelengths longer than 1 cm at cosmic background temperature (≈ 3 K). This allows us, for almost all cases where natural radiation in the microwave to submillimetre wave ranges is concerned, to rewrite Eq. (5.17) and all other related equations. If we are allowed to use Eq. (5.23) for, say, the intervening atmosphere, then we may state that the spectral intensity $I_{\nu A}$ from the background is also created by a source function $S_A(\nu)$, which in its turn may be approximated by

$$S_A(\nu) \approx 2kT_A/\lambda^2\,.$$

The observed intensity $I_{\nu B}$ can also be expressed by a temperature T_B, which is called the brightness temperature. Hence Eq. (5.17) applied to microwaves reads

$$T_B = T_A e^{-\tau_0} + \int_0^{\tau_0} T e^{-\tau} d\tau \; . \tag{5.24}$$

In general T and τ will vary along the path and τ can strongly depend on the frequency of observation, if, for example, atmospheric spectral lines are involved.

Within this section we have treated unpolarized, incoherent radiation. For this type of radiation the superposition principle for wave amplitudes is not applicable, and only the superposition of wave energy is allowed. Some of the detectors in the visible and infrared are polarization-dependent, and the usual receivers in the radio- to millimetre wave range are sensitive to only one out of two orthogonal polarizations simultaneously. The polarization direction of electromagnetic radiation is in a plane perpendicular to the direction of propagation, but within this plane any polarization direction of the electric field vector is possible. Incoherent, unpolarized radiation is characterized by a uniform distribution of the polarization directions, i.e. the intensity is distributed over all polarization directions equally. Referring the reader to Chapter 1.3, we can state that the total intensity in a beam is composed of two equal portions in two orthogonal − say the horizontal (h) and the vertical (v) − polarizations.

Measuring in one of these polarizations, we consequently receive only one half of the total intensity and we determine only one half of the source function or of the emission coefficient. For example, in the case of blackbody radiation by an optically thick source medium (5.18) will yield

$$I_{vBh} = I_{vBv} = \frac{S(v)}{2} = \frac{hv^3}{c^2} \left[\exp\left(\frac{hv}{kT}\right) - 1 \right]^{-1} , \tag{5.25}$$

and for the cases where the Rayleigh-Jeans approximation is permitted

$$I_{vBh} = I_{vBv} = \frac{kT}{\lambda^2} \; . \tag{5.26}$$

It is important here to note that the temperature of the medium is not cut into halves, one half for each polarization, when dealing with Eq. (5.26), but that the radiated intensity, according to Eq. (5.23), at a given temperature is cut into halves, one half in each polarization.

5.2 Kirchhoff's Law and Radiometry

If we were allowed to suppose a blackbody to be "black" at all wavelengths, then all impinging radiation would be perfectly absorbed without any fraction being reflected and, on the other hand, the emission of radiation due to the temperature of this body would exactly obey Planck's law [Eq. (5.16)]. For this ideal case the integration of the spectral radiance over the whole spectrum and over the hemisphere into which the radiation is emitted can be simply performed: first we integrate over the spectrum, and for this purpose we introduce a new variable $u = \dfrac{h\nu}{kT}$. With this the integration reads

$$\bar{S}_b \equiv \int_0^\infty S_b(\nu, T)\,d\nu = \frac{2(kT)^4}{c^2 h^3} \int_0^\infty u^3 [e^u - 1]^{-1}\,du \ . \tag{5.27}$$

Because of the somewhat obscure behaviour of the integrand at the integration limits, we make use of the fact that the expression in the brackets is the sum of a geometric series.

$$[e^u - 1]^{-1} = e^{-u}[1 - e^{-u}]^{-1} = \sum_{m=1}^\infty e^{-mu} \ .$$

With this the integral can be separately evaluated for each term of the series and summed up

$$\int_0^\infty u^3 [e^u - 1]^{-1}\,du = \sum_{m=1}^\infty \int_0^\infty u^3 e^{-mu}\,du = \sum_{m=1}^\infty \frac{6}{m^4} \ . \tag{5.28}$$

This result is according to standard tables of definite integrals (Groebner and Hofreiter 1966, Part 2, p. 139). The series of $1/m^4$ converges very quickly and the sum on the right-hand side of Eq. (5.28) yields the value 6.494. Using this in Eq. (5.27) we get for the frequency-integrated radiance of a blackbody surface

$$\bar{S}_b = \sigma_{SB} T^4 / \pi \tag{5.29}$$

in Watt per square metre per steradian and all constants are collected to $2\pi \times 6.494\, k^4/c^2 h^3 = 5.67 \times 10^{-8}\,\text{W m}^{-2}\text{K}^{-4} \equiv \sigma_{SB}$, the Stefan/Boltzmann constant.

For the integration over the hemisphere we assume that the emission from a plane radiating surface into the hemisphere is proportional to the projected surface area ($\bar{S}_b \cos \Theta$). Integration over the hemisphere at unit distance gives the radiated power per unit surface area of the radiating body

$$\bar{\bar{S}}_b \equiv \int_0^{2\pi} \int_0^{\pi/2} \bar{S}_b \cos \Theta \sin \Theta\,d\Theta\,d\varphi = \bar{S}_b \pi = \sigma_{SB} T^4 \ . \tag{5.30}$$

From Eqs. (5.29) and (5.30) we see that the total radiated power of a blackbody surface is proportional to the 4th power of the temperature. For a good approximation of Eq. (5.27) the integration over the frequency need not be carried out from zero to infinity, but only over that frequency range which covers the most significant orders of magnitude of the spectral radiance curve (Fig. 2.2) for the corresponding temperature. Nevertheless, even within such a reduced spectrum (at least three orders of magnitude in frequency are needed for a good approximation), only few materials in nature will behave like a blackbody (perfect absorber) or like a greybody (constant value of the albedo), according to our considerations in Chapters 2 and 3 of this text. Also

− partly for the same reason − no detector exists that is equally sensitive over this wide frequency range to measure the radiation emitted by such an ideal medium. Therefore we shall limit the following discussion to frequency bands not wider than the usual detectors can cover and a medium with an albedo $A < 1$ will be called "grey" only within these narrow frequency bands.

Despite these restrictions, let us consider a somewhat idealized experiment (Fig. 5.4). We take the spherical interior of a massive body, which is cut into two halves. The mass of each of the two half-bodies should be sufficient to prevent temperature changes on a timescale comparable to or shorter than the measurement time. The two hemispherical interior surfaces are assumed to present different albedos. We take the left one as black with an absorption factor (absorptivity) $a_b = 1$ and an emission factor (emissivity) e_b, which is still to be determined. These factors have the meaning, respectively, of the efficiency with which a surface is able to absorb any incident radiation, and with which it is able to emit all radiation available according to its temperature.

The hemisphere on the right-hand side is assumed to present a greybody surface, i.e. with an absorptivity $a < 1$ and an emissivity e. Both hemispheres are supposed to be at the same temperature T, and this thermodynamic balance is maintained if no external disturbances are applied. Therefore on the wall of each hemisphere the absorbed radiation must be balanced by the amount of emitted radiation. Because of equal temperatures, the respective Planck functions are equal, and we can omit them in computing the radiation balance. Hence on the right hemisphere the absorbed radiation is proportional to $e_b a$ and must be balanced by the emitted radiation, which is e times the same factor of proportionality (Planck function).

$$e_b a = e . \tag{5.31}$$

On the left hemisphere the incident radiation is proportional to the sum of what is emitted directly by the right hemisphere e and the reflected portion of what is incident onto the right hemisphere and not absorbed at the grey surface $e_b(1-a)$. This must be balanced by the emitted radiation, proportional to e_b.

$$e + e_b(1-a) = e_b . \tag{5.32}$$

Equations (5.31) and (5.32) are both completely equivalent and give the result

$$\frac{e}{a} = e_b . \tag{5.33}$$

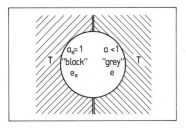

Fig. 5.4. Kirchhoff's law. Two halves of a hollow ball, one with a black, one with a grey surface

We may take arbitrarily $e_b = 1$ and Kirchhoff's law can be formulated as follows:

The ratio of emissivity and absorptivity of any material is constant and equal to unity for any given temperature. This means, of course, that the emitted spectral radiance of a grey surface is $eS_b(v, T)$, where S_b is the Planck function of a supposed blackbody at the same temperature T.

Two practical consequences can be derived from this result: first, as has been applied already in Eq. (5.32), for any opaque medium the absorptivity can be used to define a reflectivity by

$$r = 1 - a \ . \tag{5.34}$$

Second, the emissivity is numerically equal to the absorptivity

$$e = a \ , \tag{5.35}$$

and can be substituted where appropriate. If the medium is not opaque but partly transparent, the fractional transmission expressed in terms of the radiation transfer Eq. (5.14) can be directly applied only for non-reflecting interfaces. If reflection is present, the more general relation for the transmissivity,

$$t = 1 - a - r \ , \tag{5.36}$$

has to be used where the angular and polarization dependences of a and r determine the behaviour of t in this respect.

The reader will remember the statements concerning the polarization-dependent reflection of radiation at a plane interface discussed in Chapter 1.3. The Fresnel formulae [Eq. (1.52)] give the reflectivities for horizontally and vertically polarized components. Imagine now two half-spaces separated by a plane interface (Fig. 5.5).

Radiation emitted by medium 1 is partly reflected specularly, when hitting the interface, but additionally part of the radiation created by medium 2 can be emitted through the interface into medium 1. In general, this will be true not only for the whole upper hemisphere, but the same consideration can be applied to the lower one. However, for the sake of simplicity, we regard the situation for only one zenith angle Θ when trying to extend Eqs. (5.34) and (5.35) to determine the respective angular and polarization dependences of emissivity and absorptivity. Knowing that each of two orthogonal polarizations contains half of the spectral radiance available from an incoherently radiating source, the spectral intensity reflected at the interface will be

$$I_{1r}(\Theta) = r_h(\Theta) \frac{S(v, T_1)}{2} + r_v(\Theta) \frac{S(v, T_1)}{2} \ , \tag{5.37}$$

where the angular and polarization dependence of the reflectivity is given by Eq. (1.52). The spectral intensity emitted by medium 2 can be written with the, yet unknown, emissivities for the two polarizations at zenith angle Θ

$$I_{2e}(\Theta) = e_h(\Theta) \frac{S(v, T_2)}{2} + e_v(\Theta) \frac{S(v, T_2)}{2} \ . \tag{5.38}$$

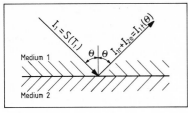

Fig. 5.5. The contributions of emitted and reflected radiation to the intensity emanating from an interface

If the two half-spaces are sufficiently extended and sufficiently independent of external radiation sources, so that we may assume thermodynamic equilibrium can be achieved and can persist for a time period longer than the measurement time, we can use $T_1 = T_2$ and $S(v, T_1) = S(v, T_2)$ in Eqs. (5.37) and (5.38). But this means that for each polarization separately we may write

$$I_{1rh}(\Theta) + I_{2eh}(\Theta) = [r_h(\Theta) + e_h(\Theta)]\frac{S(T_1)}{2}$$

$$I_{1rv}(\Theta) + I_{2ev}(\Theta) = [r_v(\Theta) + e_v(\Theta)]\frac{S(T_1)}{2} \tag{5.39}$$

(we suppress the frequency dependence of the Planck function and the subscript b for blackbody). The sum of these two lines must be

$$I_{1r}(\Theta) + I_{2e}(\Theta) = S(T_1) \ .$$

This is only possible, if

$$e_h(\Theta) = 1 - r_h(\Theta)$$

$$e_v(\Theta) = 1 - r_v(\Theta) \ . \tag{5.40}$$

The same type of reasoning can easily be made for the absorptivity, but it will suffice to state that by Eq. (5.33) or (5.35) the angular and the polarization dependences of the absorptivity are identical to those of the emissivity.

With Eq. (5.40), we can now write down the total spectral intensity I_{1t} emanating from the interface at zenith angle for horizontal and vertical polarization

$$I_{1th}(\Theta) = r_h(\Theta)\frac{S(T_1)}{2} + e_h(\Theta)\frac{S(T_2)}{2} = \frac{S(T_1)}{2} + e_h(\Theta)\frac{S(T_2) - S(T_1)}{2}$$

$$I_{1tv}(\Theta) = r_v(\Theta)\frac{S(T_1)}{2} + e_v(\Theta)\frac{S(T_2)}{2} = \frac{S(T_1)}{2} + e_v(\Theta)\frac{S(T_2) - S(T_1)}{2} \ , \tag{5.41}$$

where again different temperatures of media 1 and 2 (e.g. air and ocean) are assumed.

In Chapter 4.3 we discussed in some detail the diffuse scattering of radiation at rough surfaces. Instead of the reflectivity of a plane surface for a wave arriving from a distinct direction Θ (in Fig. 5.5), determined by the Fresnel formulae, we derived an integral [Eq. (4.61)], by which radiation incident from all directions of the (upper) half-space and partly scattered into a given direction Θ_0, φ_0 was taken into account. The result there was the albedo $r_i(\Theta_0, \varphi_0)$, where the subscript i stands for any choice of polarization. We concluded there that − for the given observing direction Θ_0, φ_0 − all radiation from the upper hemisphere, which is not reflected by surface scattering, is absorbed by the medium below the surface (4.62). According to the above considerations, it is obvious that the emissivity of any rough or plane surface into the direction Θ_0, φ_0 is related to the albedo viewed from the same direction by

$$e_i(\Theta_0, \varphi_0) = 1 - r_i(\Theta_0, \varphi_0) \ . \tag{5.42}$$

Let us regard the Lambertian surface as an example. For this case we remember, [see Eqs. (4.63) and (4.64)], that

$$\gamma_{ii}(\Theta_0, \varphi_0; \Theta_s, \varphi_s) + \gamma_{ji}(\Theta_0, \varphi_0; \Theta_s, \varphi_s) = \gamma_0 \cos \Theta_s \ ,$$

where Θ_s is now the running zenith angle coordinate instead of Θ_0. This, introduced into Eq. (4.61), yields the albedo

$$r_i(\Theta_0, \varphi_0) = \frac{1}{4\pi} \int_0^{2\pi} \int_0^{\pi/2} \gamma_0 \cos \Theta_s \sin \Theta_s \, d\Theta_s \, d\varphi_s = \frac{\gamma_0}{4} \tag{5.43}$$

and the emissivity $e_i(\Theta_0, \varphi_0) = 1 - \dfrac{\gamma_0}{4}$ is one minus one quarter the scattering coefficient at nadir. Our somewhat oversimplified consideration of the facet model in Chapter 4.3, resulting in a "Lambertian" behaviour [Eq. (4.76)], would now yield $r_i(\Theta_0, \varphi_0) = \dfrac{r_\perp}{4}$, where $r_\perp$ is given by (1.51). This result has a very limited range of validity in terms of incidence angle (Θ_0 small) and of surface modelling.

The relations between albedo and emissivity hold not only for "hard" surfaces, i.e. abrupt changes of the index of refraction within a layer much thinner than a wavelength, but also for transition layers many wavelengths thick, where volume scattering dominates the reflection properties and determines the albedo. This means also that any scattering medium (e.g. the atmospheric molecules, clouds, precipitation, the snow cover and even vegetation) for which the albedo can be defined according to Eq. (4.37), obeys the relations (5.42) and (5.35). However, depending on the particular choice of observing wavelength, these media may or may not be opaque. In the cases where they are transparent, the radiation from the background sources through the layer has to be considered by applying the equation of radiative transfer for absorbing and scattering media.

The computation of this general situation is rather complex, therefore we will consider here one example with merely the absorbing atmosphere. One can treat this situation for the relatively simple case of microwave radiometry of the earth's surface from an air-borne platform at height z_B. This case is simple, because the Rayleigh-Jeans approximation (5.23) allows computation with brightness temperatures instead of Planck functions of media at different temperatures.

In Fig. 5.6 the geometric configuration and the different contributions to the resulting brightness temperature T_B are sketched. We assume that the sky does not contribute at all, the only contribution to the downwelling radiation is due to the atmosphere, and the frequency of observation is chosen far from a spectral line, therefore no significant frequency dependence of the spectral intensity is expected within the bandwidth of the receiver. The surface is assumed rough and the subsurface medium is opaque. The further simplification of a plane and homogeneous atmosphere allows us to substitute for the range of the path by $z/\cos \Theta$, where z is the height aboveground and Θ is the zenith angle. The downwelling radiation of the atmosphere, free of scattering, expressed as apparent temperature at the ground is therefore, according to Eq. (5.24)

$$T_d(\Theta) = \frac{1}{\cos \Theta} \int_0^\infty \kappa_a(z)\, T(z) \exp\left[\frac{-1}{\cos \Theta} \int_0^z \kappa_a(z')\, dz'\right] dz \;, \qquad (5.44)$$

where $T(z)$ is the height-dependent physical temperature of the atmosphere. This radiation will be scattered at the ground. The scattering has to be taken into account by integrating the product of the scattering coefficient of the surface and the zenith angle-dependent incident radiation over the hemisphere from which this radiation arrives.

For the sake of simple analysis, let the ground behave as a Lambertian rough surface ($\gamma_0 \cos \Theta$) and the atmosphere be a homogeneous (constant κ_a) and isothermal (constant T) layer of effective height z_0 (corresponding to an optical thickness $\tau_0 = \kappa_a z$ at zenith). The rough surface causes contributions of the downwelling radiation from all directions (upper half space) to be scattered towards the observer. Integration over these contributions involves an exponential integral, $E_3(\tau_0)$, in order to find the atmospheric radiation scattered to Θ_0, φ_0

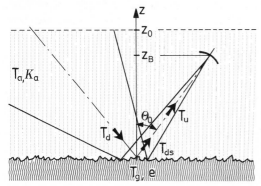

Fig. 5.6. On the contributions to the brightness temperature received by an air- (or space-) borne radiometer

$$T_{ds}(\Theta_0, \varphi_0) = \frac{T}{4\pi} \int_0^{2\pi} \int_0^{\pi/2} \gamma_0 \cos\Theta [1 - \exp(-\tau_0/\cos\Theta)] \sin\Theta \, d\Theta \, d\varphi$$

$$= T\frac{\gamma_0}{4} [1 - 2E_3(\tau_0)] \; . \tag{5.45}$$

The exponential integral $E_3(\tau_0)$ is defined and tabulated by, e.g. Abramowitz and Stegun (1970, pp. 228–251). It assumes the values 0.5, 0.2216, 0.1097, 0.0301 for $\tau_0 = 0, 0.5, 1$, and 2, respectively.

To this atmospheric contribution scattered at the surface we have to add the radiation emitted by the ground towards Θ_0, φ_0. Usually this is the important contribution to be measured,

$$T_{eg}(\Theta_0, \varphi_0) = e(\Theta_0, \varphi_0) T_g = \left(1 - \frac{\gamma_0}{4}\right) T_g \; , \tag{5.46}$$

where T_g is the physical temperature of the ground. In the last step we have to consider the attenuation of the radiation temperatures T_{ds} and T_{eg} due to the atmospheric absorption between ground and sensor and the contribution by the upwelling radiation created in the atmosphere itself. Thus we have to apply once more the equation of radiative transfer (5.24) for finding the resulting brightness temperature available to the sensor

$$T_B(\Theta_0, \varphi_0) = (T_{ds} + T_{eg}) e^{-\tau_B} + T_u \; , \tag{5.47}$$

where

$$\tau_B = \frac{1}{\cos\Theta_0} \int_{z_B}^0 \kappa_a(z) \, dz$$

is the optical thickness of the atmosphere between the observed spot at the ground and the observing platform,

$$T_u = \frac{T}{\cos\Theta_0} \int_{z_B}^0 \kappa_a(z) \exp\left[\frac{-1}{\cos\Theta_0} \int_{z_B}^z \kappa_a(z') \, dz'\right] dz \tag{5.48}$$

is the brightness temperature due to the upwelling radiation from the atmosphere. We see now that for radiometrical observation of the ground, it is important to use a wavelength range where absorption and scattering by the atmosphere are very low, otherwise the desired information T_{eg} is hidden by a strong attenuation $e^{-\tau_B}$ and by strong contributions T_{ds} and T_u of the atmosphere.

So far we have used four approximations: the sky background contribution (≈ 3 K cosmic radiation) has been neglected, the atmosphere has been assumed to be purely absorbing and free of scattering, the ground has been supposed to be a Lambertian-type surface scatterer, and finally the atmosphere has been assumed to be one homogeneous and isothermal layer. Therefore the integrals in Eq. (5.48) can be replaced by the simplified integration (5.18) of the equation of radiative transfer. From (5.47) follows the resulting equation for the brightness temperature T_B at the sensor

$$T_B(\Theta_0, \varphi_0) = \left\{ \frac{\gamma_0}{4} T [1 - 2E_3(\tau_0)] + \left(1 - \frac{\gamma_0}{4}\right) T_g \right\} e^{-\tau_B} + T(1 - e^{-\tau_B}) ,$$

(5.49)

where T is the physical temperature throughout the whole atmospheric layer, and τ_0 the optical thickness of the whole atmosphere for zenith viewing. This equation can, of course, only serve for a rough estimation, e.g. of the adverse effect of the atmosphere on the relation $T_B(T_{eg})$, because of the coarseness of our approximations. Despite all reservations against (5.49), this approach can be applied for any model of an attenuating atmosphere and for any kind of ground surface. Then, however, the integrals (5.44), (5.48) and, depending on the surface model, an integral corresponding to (5.45) have to be computed.

For ground-based down-looking sensors, the last term of Eq. (5.49) vanishes due to the negligible distance between ground and sensor ($\tau_B \to 0$), but on the other hand the expression in brackets will be received unattenuated.

So far we have tacitly assumed that this thermal radiation can be measured easily. However, in most cases the sensors are at about the same temperature as the objects, and one has not only to determine the radiance (brightness temperature) of the object, but also differences in radiance between neighbouring objects or objects and background. Moreover the "signal" contained in the radiation is, of course, only "noise" of higher or lower intensity and is indistinguishable in spectral character from the noise produced in the detectors, which usually has much higher intensity. We shall therefore devote a few paragraphs to elucidating the ideas of the radiometric measurement of thermal radiation, and we shall use the concept of microwave radiometry for this purpose.

Figure 5.7 presents schematically a microwave receiver together with an antenna surrounded by a homogeneously radiating environment. We suppose that the impedances of an effective receiver input resistance R, of the bandpass filter F, which is assumed lossless within a radio-frequency bandwidth $\Delta \nu_R$, and of the antenna, are matched to the environment which can be represented by a cavity with a perfectly absorbing inner wall w, enclosing the

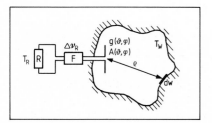

Fig. 5.7. Schematic arrangement for the definition of the antenna temperature of a microwave radiometer

antenna. The gain function of the antenna $g(\vartheta, \varphi)$ (see Chap. 1.2), can be normalized by integrating the dimensionless power gain function

$$\frac{P(\vartheta, \varphi)}{P_{tot}/4\pi} \equiv |\bar{g}(\vartheta, \varphi)|^2 \tag{5.50}$$

over all space angles

$$\frac{1}{4\pi} \int_{4\pi} |\bar{g}(\vartheta, \varphi)|^2 d\Omega = 1 \ . \tag{5.51}$$

In this definition P_{tot} is the total power emitted from an aperture and $P(\vartheta, \varphi)$ is the power per unit solid angle radiated in direction ϑ, φ. According to Nyquist, the noise power generated by the effective resistance R of the receiver at temperature T_R is given in the microwave and radio frequency regime by

$$P_R = kT_R \Delta v_R \ , \tag{5.52}$$

where k is Boltzmann's constant.

This Nyquist approximation of the noise power formula has the same range of validity as the Rayleigh-Jeans approximation of the spectral radiance, namely $v/T < 10^{10}$ Hz/K. The surface element dw of the cavity wall receives a fraction of P_R, which is determined by the normalized power gain function and the solid angle subtended by dw, when viewed from the antenna at a distance ρ,

$$dP_R = kT_R \Delta v_R \frac{|\bar{g}(S, \varphi)|^2}{4\pi} \frac{dw}{\rho^2} \ .$$

The surface element dw itself emits radiation due to its temperature T_w according to Eq. (5.23). The solid angle, which is subtended by the effective antenna area $A(\vartheta, \varphi)$, when viewed from dw is $A(\vartheta, \varphi)/\rho^2$. The amount of radiative power received by the antenna and absorbed in the resistor R is

$$dP_w = \frac{kT_w}{\lambda^2} \Delta v_R \frac{A(\vartheta, \varphi)}{\rho^2} dw \ ,$$

where a factor of $1/2$ – as in (5.26) – takes into account the fact that the antenna is sensitive to only one direction of polarization. The relation between power gain function and effective antenna area follows from Eqs. (5.50) and (1.34), it reads

$$|\bar{g}(\vartheta, \varphi)|^2 = 4\pi A(\vartheta, \varphi)/\lambda^2 \ . \tag{5.53}$$

This is well established in antenna theory and leads to the maximum antenna gain for $\vartheta = 0$, a direction for which $A(0,0) = A_{e0}$, the effective antenna area being the product of the geometric antenna area A and an efficiency factor η, usually between 0.5 and 1. This relation has already been used in Eq. (4.52) without proof, and the term gain (G) has been used for $|\bar{g}(0,0)|^2$, the maximum power gain in the principal direction of the main beam.

As soon as thermodynamic equilibrium between the effective input resistance of the receiver and the blackbody inner surface of the cavity is achieved, there will be no net exchange of radiation power, $dP_w = dP_R$, and from this follows $T_w = T_R$.

We may generalize this result to surfaces of any emissivity: an antenna pointing towards a surface at a brightness temperature T_w receives the same amount of power as a resistor at the input of the receiver would be able to generate, if it was at the same temperature T_w. This equivalence of brightness temperature of the radiating environment and input noise temperature received by the radiometer via the antenna allows us to adopt the concept of antenna temperature instead of observed radiance.

Let us assume there is not one continuous, homogeneously radiating environment surrounding the antenna, but a distribution of different brightness temperatures, within its beamwidth. Then the antenna temperature will be a combination of these different brightness temperatures, all weighted by the antenna power gain function, and determined by the sizes and the angular dependence of the emissivities of the various areas.

If the position of a source area (with, e.g. radiation temperature T_0) is localized at ϑ_0 and φ_0 with respect to the main beam direction, then the received power per unit bandwidth in polarization i is given by the convolution of the normalized power gain function $|\bar{g}_i(\vartheta, \varphi)|^2$ and angular distribution of the spectral radiances $S_{vi}(\vartheta, \varphi)$ (e.g. $S_{vi} = kT_0/\lambda^2$)

$$P_{vi} = A_e(\vartheta_0, \varphi_0) \int_{4\pi} S_{vi}(\vartheta_0 - \vartheta, \varphi_0 - \varphi)\,|\bar{g}_i(\vartheta, \varphi)|^2 d\Omega \qquad (5.54)$$

in Watt per Hertz. $A_e(\vartheta_0, \varphi_0)$ is the effective area of the sensor, e.g. the antenna area, with respect to the source position and according to the definitions (5.50) and (5.53).

For a microwave receiver the angular sizes of objects on the ground, when viewed from air-borne or space-borne platforms, are usually much smaller than the angular resolution of the antennas. We can simplify the considerations by assuming a constant sensitivity of the antenna over a solid angle Ω_A defined by

$$\Omega_A \equiv \int \frac{|\bar{g}(\vartheta, \varphi)|^2}{|\bar{g}_{max}|^2} d\Omega = \frac{\lambda^2}{A_{e0}} \qquad (5.55)$$

at a wavelength λ and with an effective antenna area A_{e0} in the principal direction of the main beam where $\bar{g}(0,0) = \bar{g}_{max}$ is assumed. The comparison of Eq. (5.55) with (1.36) for the one-dimensional case is evident. At a distance ρ (assumed to be in the far-field zone, where the Fraunhofer diffraction theory is valid) the main beam covers a cross-section area of

$$A = \Omega_A \rho^2 \approx \lambda^2 \rho^2 / A_{e0} . \qquad (5.56)$$

For example, the beam of an antenna of $A_{e0} = 1 \text{ m}^2$, used at $\lambda = 1$ cm, covers a cross-section of about 100 m^2 at a distance of 1 km.

Now let us consider the ability of a microwave sensor to detect a remote object at distance ρ presenting a brightness temperature T_0 different from the surrounding T_s by ΔT_0. The cross-sectional area of the object A_0 subtends a solid angle

$$\Omega_0 = A_0/\rho^2 , \qquad (5.57)$$

and we assume $\Omega_0 < \Omega_A$.

When neglecting the effect of the atmosphere, the antenna temperature as a function of the brightness temperatures and of the ratio of solid angles can be given by

$$T_A \approx \frac{\Omega_0}{\Omega_A} T_0 + \left(1 - \frac{\Omega_0}{\Omega_A}\right) T_s . \qquad (5.58)$$

Sweeping the antenna beam over the object embedded in a homogeneous surrounding, the change of the antenna temperature will be

$$\Delta T_A = \frac{\Omega_0}{\Omega_A} \Delta T_0 , \qquad (5.59)$$

with $\Delta T_0 = T_0 - T_s$.

Thus a small ratio Ω_0/Ω_A reduces the observable increment of brightness temperatures proportionally. If the observation frequency is chosen in a range where the atmospheric absorption and emission has to be taken into account, the change of antenna temperature due to an object becomes even smaller. We can consider for this purpose an approximate formulation of Eq. (5.49). We introduce the emissivities e_0 of the object and e_s of the surroundings and approximate the downwelling atmospheric radiation by the vertical component $(1 - e^{-\tau_0}) T$. This approximation approaches the values $[1 - 2E_3(\tau_0)] T$ of Eq. (5.45) for increasing τ_0. With this we write $e_s T_s$ instead of $\left(1 - \frac{\gamma_0}{4}\right) T_g$ in Eq. (5.49), when the homogeneous surrounding scenery is viewed, and we write $(1 - \Omega_0/\Omega_A) e_s T_s + \Omega_0/\Omega_A e_0 T_0$ when the object – embedded in the surroundings – is within the beam. In most practical situations it will be permissible to assume that both the physical temperatures of the object T_0 and of the surrounding ground T_s are identical. The difference of the antenna temperatures of both observations,

$$\Delta T_A = \frac{\Omega_0}{\Omega_A} (e_0 - e_s) [T_s - (1 - e^{-\tau_0}) T] e^{-\tau_B} , \qquad (5.60)$$

is the resulting information at the input of the receiver. The information on the object is now contained in $(e_0 - e_s)$ or, expressed as brightness temperatures, in $(e_0 - e_s) T_s = \Delta T_0$.

The difference of the antenna temperatures, when the antenna sweeps over the object ΔT_A, is the "signal" received by the antenna. According to

Eq. (5.52), this antenna temperature change presents a change of microwave power,

$$\varDelta P_A = k \varDelta v_R \varDelta T_A \; , \tag{5.61}$$

to the input of the receiver. This small power (for a $\varDelta T_A = 1$ K and a bandwidth $\varDelta v_R = 1$ MHz of the receiver, $\varDelta P_A$ is only about 1.4×10^{-17} Watt) has to compete with the noise power produced within the receiver itself. This is usually two or three orders of magnitude larger.

The measurement of a tiny noise power as the "signal" superimposed on a huge noise background of the same spectral properties is a typical problem of statistics; it can be solved only by a large number of independent measurements.

According to the concept of a one-dimensional "random walk" with no correlation between successive steps (each of which can either be $+1$ or -1), the mean square distance covered after n steps is given by $\langle x^2 \rangle = nl^2$ (e.g. MacDonald 1962, pp. 12–15).

If we assume that v steps occur per second, we have

$$n = vt \tag{5.62}$$

in a time interval t. In an observation involving n individual random events, the typical magnitude of displacement is the root mean square distance

$$\langle x^2 \rangle^{\frac{1}{2}} = \sqrt{n}\, l \; . \tag{5.63}$$

When measuring a given length with a ruler, the error in each measurement may be assumed to be plus or minus one unit on the ruler (say 1 mm). The accuracy of the measurement can obviously be increased by repeated measurements. However, in that case the sum of the errors (5.63) is not of relevance, but only the relative error is important. This can be defined by dividing $\langle x^2 \rangle^{1/2}$ by the maximum possible error nl after n measurements. Hence

$$\frac{\langle x^2 \rangle^{1/2}}{nl} = \frac{1}{\sqrt{n}} \; . \tag{5.64}$$

A radio frequency receiver of limited bandwidth $\varDelta v_R$ at the input stage can only receive a maximum of $\varDelta v_R$ fluctuations per second [cf. Chap. 1.2, in particular the discussion after formula (1.30)]. In such a receiver we can also limit the duration of measurement, i.e. the duration of summing up individual fluctuations by a detector with a "memory" of finite length. This can be realized by a simple RC circuit; its reaction time determines the duration of the summing-up process and is called integration time $\varDelta t$. We can identify in (5.62) t with $\varDelta t$ and v with the radio-frequency (predetection) bandwidth $\varDelta v_R$. From Eq. (5.64) it follows that a radiometer, i.e. a receiver of microwave thermal radiation, can detect an antenna temperature with a relative sensitivity of

$$\frac{(\varDelta T_A)_{min}}{T_{RN}} = \frac{1}{\sqrt{\varDelta v_R \varDelta t}} \; , \tag{5.65}$$

where $T_{RN} = T_A + T_N$ is the effective noise temperature at the receiver input, i.e. the sum of the radiation received via the antenna and the noise produced within the receiver itself. The limit of detectability can be assumed to be a signal-to-noise ratio of unity at the detector. For a receiver with an RF input bandwidth $\Delta v_R = 1$ MHz and a noise temperature $T_N = 700$ K, changes of $\Delta T_A = 1$ K superimposed on a background of $T_A = 300$ K can just be detected, if the integration time is at least 1 s. Consequences of this slow reaction time and the trade-offs between Δt, Δv_R, T_{RN} and antenna beam-width Ω_A for detecting isolated objects, in particular by scanning air- or space-borne radiometers, have been discussed elsewhere (Schanda 1971). For the determination of the absolute value of the object brightness temperature, for the detection of narrow band spectral features and for achieving high angular resolution, different concepts of sophisticated radiometers have been developed (e.g. Tiuri 1966, where the performances are also analyzed).

So far we have discussed only some aspects of microwave radiometry. Let us conclude this section with a few remarks on radiometry in the infrared. There exists a great diversity of infrared radiation detectors, dependent on the specific wavelength range, on atmospheric or surface applications, and on the use in imaging or non-imaging systems. Detailed presentations of modern remote sensing infrared detectors are given by Robinson and de Witt (1983) and by Norwood and Lansing (1983), while R. A. Smith et al. (1968) treat many important fundamentals and the more classical types of detectors in a comprehensive way.

In broadband infrared detection one has to keep in mind that the Stefan-Boltzmann law (5.29) has to be applied, when dealing with the frequency-integrated radiance, and that there is a much stronger dependence on temperature T than on emissivity e of a non-blackbody due to

$$\bar{S}(T) = \sigma_{SB} \, e(T) \, T^4 / \pi \ , \tag{5.66}$$

where the spectral mean of the emissivity e can still be a function of the temperature.

In narrowband applications, e.g. the detection of atmospheric spectral lines, however, both the emissivity and, to a lesser extent, the blackbody radiance are strongly frequency-dependent, yielding a resultant spectral radiance

$$S(v, T) = e(v, T) \frac{2hv^3}{c^2} \left[\exp\left(\frac{hv}{kT}\right) - 1 \right]^{-1} \ . \tag{5.67}$$

This cannot be integrated analytically with respect to the frequency and in general does not give a total radiated power proportional to T^4.

The performance of infrared detectors is usually described by a few figures of merit (Hudson 1969). When the incident radiation power (intensity I times the sensitive area of the detector A_d) are related to the root mean square of the output voltage V_0 one defines the responsivity in volts per Watt

$$R_0 = \frac{V_0}{I A_d} \ . \tag{5.68}$$

The responsive time constant Δt_R of the detector is defined by the time it takes for the detector output to reach $\left(1 - \dfrac{1}{e}\right) \approx 0.63$-fold of the final value after a sudden change of the irradiance, assuming a simple exponential law. This can be used to relate the responsivity as a function of the chopping frequency ω_c to Δt_R and to R_0, the low frequency value of the responsivity,

$$R(\omega_c) = R_0 [1 + \omega_c^2 \Delta t_R^2]^{-1/2} \ . \tag{5.69}$$

The responsivity is a very useful dynamic figure of merit, but when the ultimate sensitivity is required, the concept of the noise equivalent power (NEP) is in common use. The NEP is the radiant flux (incident power) which gives an output signal equal to that of the detector noise. Since it is difficult to measure output signals as well as radiant fluxes at those low levels where the signal to noise ratio is unity, it is customary to perform the measurement at a sufficiently high signal level $V_0 > V_N$ and to calculate

$$\text{NEP} = I A_d V_N / V_0 = V_N / R_0 \ , \tag{5.70}$$

where V_N is the RMS value of the noise voltage and V_0 the output voltage due to the signal. The unit of NEP is Watt. Some authors prefer to use the reciprocal of NEP and call this quantity the detectivity $D = 1/\text{NEP}$. When detectors of different manufacturers are to be compared, a standardized measurement has to be observed, e.g. temperature of the detector, wavelength of the incident radiation, chopping frequency and bias current of the detector have to be applied at standardized values. The sensitive area A_d of the detector and the bandwidth of the low-frequency detector circuit Δv_d have to be taken into account by the formulation of the relevant figure of merit, the normalized detectivity

$$D^* = D(A_d \Delta v_d)^{1/2} = \frac{(A_d \Delta v_d)^{1/2}}{\text{NEP}} \ , \tag{5.71}$$

which is usually given in units of $\text{cm W}^{-1} (\text{Hz})^{1/2}$. This D^* has been widely accepted, because for a given type of detector the quantity $D\sqrt{A_d}$ has been found constant and on the assumption that the noise voltage per Hertz bandwidth is independent of frequency, $D\sqrt{\Delta v_d}$ is also constant.

5.3 Radiometric Observation of Atmospheric Parameters and the Inversion of Remotely Sensed Data

A classical remote sensing application of radiative transfer is the sounding of atmospheric parameters. On several occasions we have discussed various electromagnetic properties of atmospheric gases; e.g. the spectral lines due to transitions between different molecular or atomic states which are characteristic for each species, the spectral widths of these lines which contain information on density and/or temperature, and the strongly frequency-dependent absorption coefficient resulting from these features. All these properties can be introduced now in the equation of radiative transfer in order to demonstrate its ability to infer, from frequency-resolved remote absorption or emission measurements, the atmospheric parameters: temperature, pressure and composition. But the limitations and difficulties of this approach can also be made evident.

Figure 5.8 shows a summary of the global height distributions of various minor constituents in the stratosphere and mesosphere. If one wants to measure the presence or the height distribution of a trace constituent by radiometrically detecting its proper radiation, one obviously has to make use of the spectral characteristics of this molecule in the given environment. For example, ozone at the maximum of its relative abundance at a height of about 25 to 30 km is represented by 1 molecule in about 10^5 molecules of all other constituents.

The molecular transition lines not only allow identification, but due to large absorption coefficients near line centre, the detection of very dilute constituents in a sea of background species becomes feasible.

Let us regard the radiation upwelling from the lower levels to the top of the atmosphere and compute the resulting intensity. This is the situation of observation by a spaceborne sensor looking vertically downward. For this situation we can write the spectral radiation intensity arriving at the sensor, according to the formal integration (5.17) of equation of radiation transfer

$$I_{vs} = I_{vg}\exp\left[-\int_{z_g}^{z_s} \kappa_a(v,z)\,dz\right]$$

$$+ \int_{z_g}^{z_s} S[v, T(z)]\exp\left[-\int_{z}^{z_s} \kappa_a(v,z')\,dz'\right]\kappa_a(v,z)\,dz \ . \tag{5.72}$$

Here the subscripts s and g have the meaning of "sensor" and "ground" respectively. The important features of this equation are that the absorption coefficient $\kappa_a(v,z)$ is now a very rapidly varying function of frequency, the variation depending also on height, while

$$S[v, T(z)] = \frac{2hv^3}{c^2}\left[\exp\frac{hv}{kT(z)} - 1\right]^{-1} \tag{5.73}$$

the Planck function slowly varies with frequency, but it depends also on height, because the temperature varies with height.

When the absorption coefficient of the observed spectral line is so high that $\tau_0 \gg 1$, the first term on the right-hand side (r.h.s.) of Eq. (5.72) can be

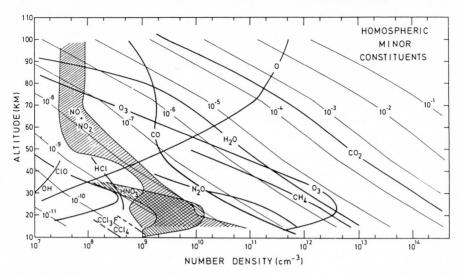

Fig. 5.8. General view of the abundance of minor constituents in the stratosphere and mesosphere. (Ackerman 1976)

neglected (opaque atmosphere). For abbreviating the second term of the r.h.s. of Eq. (5.72) we introduce the incremental transmission

$$dt_v = d\{\exp(-\int \kappa_a dz)\} = -\exp(-\int \kappa_a dz')\kappa_a dz \ , \tag{5.74}$$

where d is the differential operator acting on the spectral transmission t_v (the term in brackets).

It is often convenient to replace the height variable z by the pressure-dependent variable $y = -\ln p$, where p is the pressure in atmospheres.

With the variable y the spectral intensity received becomes

$$I_{vs} = \int_{z_g}^{z_s} S(v, T) \frac{dt_v}{dy} dy \ . \tag{5.75}$$

The intensity is the weighted average of the blackbody radiance (the Planck function), the weighting function being

$$w(y) = dt_v/dy \ . \tag{5.76}$$

The variable y is employed instead of the height z, because then the weighting function becomes more nearly independent of temperature.

The specific form of the transmission t_v and of the weighting function $w(y)$ in sensing of infrared atmospheric radiation (e.g. the 15 μm CO_2 bands for temperature sounding) depends on the specific choice of the line, the bandwidth and the exact wavelength of the sensor, i.e. on whether layers of the atmosphere at lower or higher pressures are to be sensed (Houghton et al. 1984). In any case the weighting function determines the rather thick layer from which the measured radiation originates. In the somewhat easier case of sounding by millimetre waves, we shall discuss the weighting function in more detail.

Not only is Eq. (5.72) the fundamental relation for spectrally resolved sensing of the line shape, but by integrating this equation over frequency, one can compute the contribution of this specific molecular transition to the atmosphere's energy budget from radiative processes (e.g. Houghton 1977). For integration over the frequency band of a single spectral line, the Planck function can be regarded as a constant.

Returning to the remote sensing of atmospheric parameters, let us treat the rather simple case of sounding the height distribution of the atmospheric temperature by using spectral lines in the millimetre wave range. For this purpose we select a constituent with a well-known height distribution of its relative density. Choosing the millimetre wave spectrum yields two advantages. First, it allows the use of the Rayleigh-Jeans approximation of Eq. (5.73), which means we may substitute for intensity I and source function S, respectively, the brightness temperature and the physical temperature of the atmosphere, the latter being, of course, frequency-independent. Second, it removes the dependence on the absolute number density of the constituents, which is still contained implicitly in κ_a of Eq. (5.72). The latter is because for millimetre transitions the pressure broadening of the line is the dominating effect determining the linewidth (see Fig. 2.12) and in the pressure-broadened regime, the absorption coefficient is proportional only to the (relative) volume mixing ratio of the constituent in question. Another advantage of selecting a millimetre wave transition is that the ocean surface presents a very inefficiently emitting background, reflecting a considerable fraction of the low brightness temperature of the sky. With this cold background, one can observe the atmosphere in emission down to sea surface level. The best-known height distribution of species with spectral lines in the millimetre range is of the oxygen molecule.

With the above-mentioned assumption, Eq. (5.72) is simplified to

$$T_{vs} = T_{vg} \exp\left[-\int_{z_g}^{z_s} \kappa_a(v,z)\,dz \right] + \int_{z_g}^{z_s} T(z) \underbrace{\exp\left[-\int_{z}^{z_s} \kappa_a(v,z')\,dz' \right] \kappa_a(v,z)}_{w_T(v,z)} \, dz .$$

$$(5.77)$$

Here T_{vs} and T_{vg} are the (frequency-dependent) brightness temperatures at sensor and ground level respectively, and $T(z)$ is the height-dependent physical temperature of the atmosphere. The part $w_T(v,z)$ of the last term of Eq. (5.77) is the weighting function for temperature sounding.

Figure 5.9 gives an intuitive explanation of the weighting function. The left part of the figure (a) presents schematically the pressure p as a function of height in the atmosphere. We select three different height levels and determine the line shapes of a molecular transition line (e.g. of O_2) at these three levels. The right part of the figure (b) indicates the pressure-broadened line shapes corresponding to the three height levels. If a downward-looking radiometer

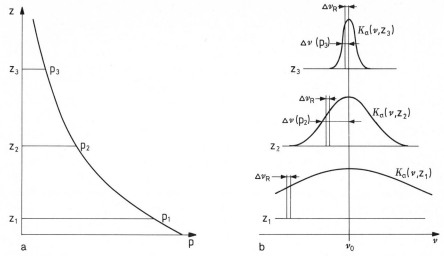

Fig. 5.9. a The atmospheric pressure as a function of the height; **b** the line shapes at three different height levels and filter positions

receives radiation through a filter whose spectral width is narrower than the width $\Delta \nu$ of the molecular line (the lower limit of which is set by the Doppler width), then the received radiation will originate from different levels dependent on the exact frequency on which the filter is set. When the filter position ν_3 is close to the centre frequency of the line (case z_3 in Fig. 5.9), the highest levels of the atmosphere ($z \gtrsim z_3$) will contribute strongly to the observed radiation, because $\kappa_a(\nu_3, z)$ is very large. However, due to this high absorption coefficient, the exponential function $\exp \left[-\int\limits_{z}^{z_s} \kappa_a(\nu_3, z') dz' \right]$,

representing the transmission of radiation from below z through the layer between z and z_s, becomes very small at rather high levels z. When the filter is shifted away from the line centre (ν_2), there is no contribution to the received radiation by layers at about z_3 and higher. The molecular line is so narrow there that through this filter position no radiation is received from these high levels. However, the lower levels at about z_2 present a sufficient width of the spectral line to contribute radiation through the filter band at this position (ν_2). The received radiation will increase due to increase of $\kappa_a(\nu_2, z)$ with decreasing z due to the corresponding widening of the spectral line, until the transmission decreases again. The result is an approximately bell-shaped weighting function for each filter position. In Fig. 5.10 the weighting functions are presented for a 12-channel radiometer at the oxygen lines between 5 and 6 mm wavelength.

The filter bandwidths are approximately adjusted to the absorption coefficients at the respective filter settings, still leaving slight differences of the

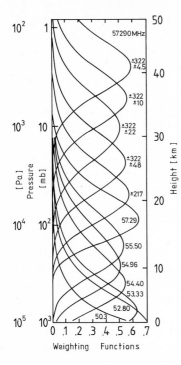

Fig. 5.10. Weighting functions for the AMSU-oxygen channels for a satellite viewing a standard atmosphere at nadir. (Staelin 1983)

sensitivities (maximum values of the weighting functions) in the different channels.

This sophisticated temperature sensor is planned as part of the Advanced Microwave Sounding Unit (AMSU) for the next generation of operational meteorological satellites starting in the late 1980's (Staelin 1983). Considerable experience has already been gained with experimental temperature sensors of more modest design since the early 1970's, e.g. with the three-channel Nimbus-5 Microwave Sounder (NEMS) (Waters et al. 1975a).

Taking half the maximum values of the weighting function to characterize the layer thickness from which most of the radiation originates, one recognizes that the temperature profile is averaged over an approximately 10-km layer for each setting of the filter.

One also recognizes that for three filter settings of AMSU the radiative contribution from the ground is significant. In these cases the first term on the r.h.s. of Eq. (5.77) cannot be neglected. For the other channels the atmosphere appears opaque. Because of the finite spectral widths of the filters, all terms of Eq. (5.77) with $\kappa_a(v,z)$ have to be multiplied by a filter function $f(v)$ and integrated over the respective transmission bands. With this supposition and according to the above discussion of microwave temperature sounding, Eq. (5.77) can be simplified to

$$T_{v_is} = \int_{z_g}^{z_s} T(z)\, w_T(v_i,z)\, dz \ , \tag{5.78}$$

where v_i stands for the respective centre frequencies of the filter, and $w_T(v_i, z)$ is the weighting function for each filter setting v_i. Obviously, in order to compute the temperature profile $T(z)$ from the measured $T_{v_i s}$ values, an inversion procedure has to be performed. That is to say that from the measurement of the line shape $T_{v_i s}$ as it evolves by radiative transport through the atmospheric layer, one is able to derive the height profile of the atmospheric temperature. Figure 5.11 shows the result of a temperature sounding with a satellite-borne radiometer having only three filter channels and also using the 5-mm oxygen absorption band. For comparison the radio sonde measurements are also given. Both the polar as well as the tropic temperature distribution between ground and 25 km are fairly well measured. Only very abrupt changes of the temperature versus height are not well observable, because of the width of the weighting function.

If not the temperature, but the unknown height distribution of an atmospheric constituent is to be measured, e.g. water vapour, ozone, carbon monoxyde or others, one can also make use of Eq. (5.77). However, the density of a species cannot be measured if the absorption coefficient in the spectral line is so high that the atmosphere appears opaque, as in that case only the temperature is observed. With an optically thin line ($\tau_0 < 1$) against a background that provides sufficient contrast (cold space or cold space reflected by the ocean), the measurement of the abundance as a function of height is equally feasible as temperature sounding. An important prerequisite

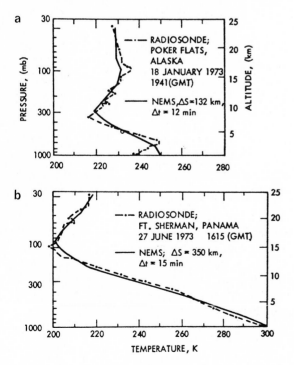

Fig. 5.11. Atmospheric temperature profiles as obtained by the inversion of the NEMS data over **a** a polar and **b** a tropical site. For comparison, radio sonde measurements are plotted. (Waters et al. 1975b)

for the following formulation is the spectroscopically established fact that the absorption coefficient at line centre $\kappa_a(v_0)$ is independent of pressure (i.e. of height) as long as the linewidth is determined by the pressure-broadening mechanism [see formulae (2.27) and (2.26) for $v = v_0$]. This allows us to factor the absorption coefficient

$$\kappa_a(v, T, z) = \kappa_{a0}(v_0, T) s(v, T, z) \cdot V(z) , \qquad (5.79)$$

where $\kappa_{a0}(v_0, T)$ is the absorption coefficient at centre frequency of the line and at standard pressure. The line shape function s is normalized by $s(v_0, T, z) = 1$ and $V(z)$ is the volume mixing ratio, i.e. the ratio of the number density of the constituent in question to the total atmosphere density at height z. With the assumption of an optically thin line, i.e. the fractional transmission $\exp\left[-\int_{z_g}^{z_s} \kappa_a dz\right] \approx 1$, the absorption by the line can be neglected.

Disregarding the effects of other atmospheric constituents on the absorption of the background radiation and on the emission within the atmosphere, we put Eq. (5.77) in the simple form of

$$T_{vs} \approx T_{vg} + \int_{z_g}^{z_s} V(z) \underbrace{T(z) \kappa_{a0}(v_0, T) s(v, T, z)}_{w_v(v, z)} dz , \qquad (5.80)$$

where $w_v(v, z)$ is the weighting function for determining the height profile of the constituent's volume mixing ratio. Equation (5.80) has to be modified for the real atmosphere: the combined effect of all other constituents would cause a transmission factor $\exp(-\tau_c)$ due to the continuum absorption (far away in the spectrum from any line frequency), to be multiplied with both terms on the r.h.s. of Eq. (5.80). There would also be an additional emission term like the last term of Eq. (5.77) due to the continuum contribution. For an observation direction against the sky a very low background radiation (a few Kelvin) can replace the radiation T_{vg} from the ground.

The first world maps of total water vapour over the oceans derived from a nadir-looking microwave radiometer (at the 22.3 GHz water vapour line) were reported by Staelin et al. (1976). Figure 5.12 presents their results in equidensity lines (g H_2O vapour cm^{-2}) over the tropical regions of the globe. Due to the weakness of this absorption line only the total water vapour over the surface (measured at line centre) could be determined. Height profiles of water vapour have so far been retrieved only from infrared multichannel space-borne radiometers via weighting functions which were derived from, e.g. the rotation band of the H_2O molecule between 20 μm and 40 μm (W. L. Smith 1970). At present, infrared sensors on board NOAA-satellites deliver multichannel data on a routine basis for the retrieval of height profiles of humidity and temperature.

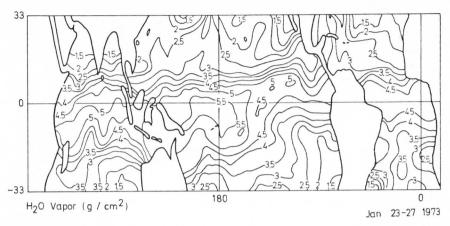

Fig. 5.12. Map of the water vapour distribution over the oceans between ± 33° latitudes obtained by the 22 GHz channel of NEMS. (Staelin et al. 1976)

All three formulations of the received radiation at a given frequency as a function of the radiation produced by the incremental layers along the height z, Eqs. (5.75), (5.77) and (5.80) can be reduced to the common form

$$g(v) = \int_{z_g}^{z_s} f(z)\, w(v,z)\, dz \ . \tag{5.81}$$

Obviously this integral equation cannot, in general, be solved analytically for $f(z)$; however, the problem of remote sensing is to solve Eq. (5.81) for the continuously variable, unknown function $f(z)$ (e.g. the height distribution of an atmospheric parameter) from the given function $g(v)$ usually measured at a finite number of discrete frequencies v_i. The weighting function $w(v,z)$ plays the role of the kernel in this integral equation. In general this integral equation is non-linear because $f(z)$ can also vary with frequency; e.g. in Eq. (5.75) we have $f(z,v) = S_v[T(z)]$ the Planck function. The problem of solving Eq. (5.81) is called an inversion problem or an inverse measurement. A considerable amount of literature exists on the inverse problem, among which the book by Twomey (1977b) is the most comprehensive and most appropriate for our purpose. We shall follow essentially his formulation in the remaining part of this section.

The numerical calculation of the continuous $f(z)$ from discrete values of $g(v_i)$ must rely on the approximation of the integral by a sum. Such approximation is known as numerical quadrature and a variety of quadrature formulae exists. The accuracy of a quadrature can be increased by an increasing number of points z_i, i.e. by the degree of subdivision within z_g and z_s. An increase of accuracy by increasing the number of measurements is limited by the smoothness of the weighting functions and by the limited precision with which the measurement can be performed.

A simple but useful quadrature procedure involves dividing the interval (z_g, z_s) by interposing n so-called quadrature points $z_1, z_2, \ldots z_n$, envisaging $f(z)$ to take the respective values $f_1, f_2, \ldots, f_n$ at those points and to behave linearly across the subintervals (Fig. 5.13).

The integral (5.81) is then approximated by summing the integrals

$$\int_{z_j}^{z_{j+1}} (A_j + B_j z) w_i(z) \, dz \tag{5.82}$$

over all the subintervals, where $w(v_i, z)$ has been contracted to $w_i(z)$. A_j and B_j are chosen to make $A_j + B_j z$ coincide with $f(z)$ at z_j and at z_{j+1}:

$$A_j + B_j z_j = f_j$$

$$A_j + B_j z_{j+1} = f_{j+1} \; .$$

From this follows

$$B_j = \frac{f_{j+1} - f_j}{z_{j+1} - z_j} \; , \quad A_j = \frac{z_{j+1} f_j - z_j f_{j+1}}{z_{j+1} - z_j} \; . \tag{5.83}$$

Thus the contribution (5.82) by the interval (z_j, z_{j+1}) can be broken up in factors of f_j and f_{j+1}

$$\frac{1}{z_{j+1} - z_j} [\{z_{j+1} \int w(z) \, dz - \int z w(z) \, dz\} f_j - \{z_j \int w(z) \, dz - \int z w(z) \, dz\} f_{j+1}] \; , \tag{5.84a}$$

where all integrals are taken between the limits z_j and z_{j+1}. The function f_j will, of course, appear again with the interval (z_{j-1}, z_j):

$$\frac{1}{z_j - z_{j-1}} [\{z_j \int w(z) \, dz - \int z w(z) \, dz\} f_{j-1} - \{z_{j-1} \int w(z) \, dz - \int z w(z) \, dz\} f_j] \; , \tag{5.84b}$$

where all integrals are taken between the limits z_{j-1} and z_j. Adding the contributions of all subintervals and arranging them according to factors of f_{j-1}, $f_j, f_{j+1}, \ldots f_n$, one obtains the quadrature formula

$$\int_{z_g}^{z_s} f(z) \, dz \approx \sum_{j=1}^{n} a_j f_j \; , \tag{5.85}$$

where the coefficients a_j follow from Eq. (5.84) as

$$a_j = \frac{1}{z_{j+1} - z_j} \left[z_{j+1} \int_{z_j}^{z_{j+1}} w(z) \, dz - \int_{z_j}^{z_{j+1}} z w(z) \, dz \right]$$

$$- \frac{1}{z_j - z_{j-1}} \left[z_{j-1} \int_{z_{j-1}}^{z_j} w(z) \, dz - \int_{z_{j-1}}^{z_j} z w(z) \, dz \right] \; .$$

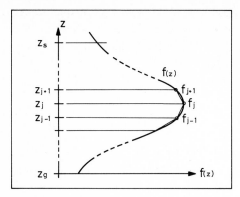

Fig. 5.13. A simple quadrature scheme

The integrals for each subinterval may be evaluated numerically or analytically. The accuracy of the quadrature depends alone on how good the linear approximation to $f(z)$ is across each interval.

We can now substitute for the integral equation (5.81) the linear approximation

$$G = \mathbb{W}F \ , \tag{5.86}$$

where G and F are vectors of dimensions m and n respectively corresponding to the m measurements $g(v_i)$ at m frequencies v_i, $(v_1, v_2 \ldots, v_m)$, and to the n values of the function $f(z_j)$ at selected height levels z_j, $(z_1, z_2, \ldots z_n)$ in the atmosphere, while $\mathbb{W}$ is a $m \times n$ matrix with elements w_{ij}, relating the j-th quadrature coefficient with the i-th kernel. The integral equation

$$g(v_i) = \int w(v_i, z) f(z) \, dz \approx \sum_{j=1}^{n} w_{ij} f(z_j) \tag{5.87}$$

is reduced to a system of linear equations

$$g(v_1) = w_1(z_1) f(z_1) + w_1(z_2) f(z_2) + \ldots + w_1(z_n) f(z_n)$$

$$g(v_2) = w_2(z_1) f(z_1) + w_2(z_2) f(z_2) + \ldots + w_2(z_n) f(z_n) \tag{5.88}$$

$$\vdots$$

$$g(v_m) = w_m(z_1) f(z_1) + w_m(z_2) f(z_2) + \ldots + w_m(z_n) f(z_n) \ ,$$

which is completely equivalent to the matrix formulation (5.86).

We have now arrived at a presentation which equals the one derived from very simple arguments in Chapter 1.1 of this text.

The numbers m and n are often large and the measured nature of g as well as the finite accuracy of quadrature formulae always cause a small error component in the linear equation (5.87). This can be expressed in the matrix formulation (5.86) as

$$\mathbb{W}F = G + \varepsilon \ , \tag{5.89}$$

where ε is an error due to the quadrature. The error in the quadrature operation is not large when the functions involved are reasonably smooth and a sufficient number of appropriately chosen values z_j are used.

Returning now to the inversion itself, one very often finds that the result of the operation

$$F' = \mathsf{W}^{-1}G = F - \mathsf{W}^{-1}\varepsilon$$

is far different from the original value of F. The small error can be amplified by the inversion if a near-singular matrix, i.e. a matrix with a least one very small eigenvalue, is involved. If W has a small eigenvalue, its inverse W^{-1} will have a large eigenvalue and inevitably will thereby contain some large elements.

A small digression here will serve to remind the reader of the definition of the eigenvalue of a matrix. If for a matrix W, containing off-diagonal terms, a vector V is found such that $\mathsf{W}V$ is a simple scalar multiple of V

$$\mathsf{W}V = \lambda V \ , \tag{5.90}$$

then λ is called the eigenvalue and V is the eigenvector. The determination of the eigenvalues follows immediately from Eq. (5.90), which, if written in full, is a homogeneous set of linear equations in the variables $v_1, v_2, \dots v_n$

$$
\begin{aligned}
(w_{11} - \lambda)v_1 + \quad & w_{12}v_2 + \dots \quad + w_{1n}v_n && = 0 \\
w_{21}v_1 + (w_{22} - \lambda)v_2 + \dots \quad & + w_{2n}v_n && = 0 \\
\vdots \quad & && \\
w_{n1}v_1 + \quad & w_{n2}v_2 + \dots \quad + (w_{nn} - \lambda)v_n && = 0 \ .
\end{aligned}
\tag{5.91}
$$

A nontrivial solution of these equations can only exist when the associated determinant vanishes. For a 2×2 matrix this condition on the determinant is

$$(w_{11} - \lambda)(w_{22} - \lambda) - w_{12}w_{21} = \lambda^2 - (w_{11} + w_{22})\lambda + w_{11}w_{22} - w_{12}w_{21} = 0 \ .$$

This quadratic equation in λ has two roots, λ_1, λ_2, which can be complex. In general the determinant of a $n \times n$ matrix is an n-order polynomial in λ possessing n roots λ_1, $\lambda_2, \dots \lambda_n$ which are the eigenvalues of the matrix.

If the direct inversion of Eq. (5.86) does fail in many applications due to error amplification, one could either look for more measurement data and use the overdetermined system of m equations to reduce the errors in the n unknowns $(n < m)$ by obtaining a least-squares solution, or search for a different method of inversion. However, the least-squares method proves to be no better than the direct inversion as far as error amplification by near-singular matrices is concerned (Twomey 1977b). One of various inversion methods is the so-called constrained linear inversion. For the solution of $f(z)$ of Eq. (5.81) with finite errors on the $g(v_i)$ measured values there will exist in the (z, f) domain a set (probably an infinite number) of functions $f(z)$. The ambiguity can only be removed by imposing an arbitrary additional condition. For example, one may ask for the smoothest $f(z)$ or the $f(z)$ with the smallest deviation from the mean. Such a condition could be formulated so as to minimize

$$|\mathbb{W}F-G|^2+\Gamma q(F) \; , \tag{5.92}$$

where $q(F)$ is some non-negative scalar measure of the deviations from smoothness of F and Γ is a parameter which can be varied from zero to infinity. The assumption of $\Gamma > 0$ admits a finite deviation from the pure least mean square minimization of $(\mathbb{W}F-G)^2$ and $\Gamma < \infty$ admits a deviation from perfect smoothness. The "goodness" of F as a possible solution of the equation $\mathbb{W}F = G$ can be defined by a fixed value of the admissible error value e such that

$$|\mathbb{W}F-G|^2 \leqslant e^2 \tag{5.93}$$

is fulfilled by a small subset of all possible vectors F. Provided e^2 is of the same order as the uncertainties of the measured G, all vectors in the subset are acceptable solutions. From this subset a unique F can be selected, which is the smoothest, as judged by the measure $q(F)$. The intermediate value of Γ produces a solution which is the result of the trade-off between deviations from smoothness and oscillatory behaviour characteristic for least-squares solutions.

The smoothness can be measured by simple quadratic combinations of the f_i. Therefore, these are quadratic forms of F, which can be written as

$$q = F^* \mathbb{H} F \; , \tag{5.94}$$

where $\mathbb{H}$ is usually a simple near-diagonal matrix and F^* is the transpose of vector F. If a vector F is expressed as a $(n \times 1)$ matrix, then its transpose F^* is a $(1 \times n)$ matrix and the product of both gives a scalar, equal to the length of F squared

$$|F|^2 = F^*F \; ,$$

a result which is also called the scalar product in geometric vector analysis. If a matrix $\mathbb{W}$ is transposed, in the resulting matrix $\mathbb{W}^*$ the elements are arranged as if the original arrangement were reflected at the main diagonal, i.e. if $\mathbb{W}$ is a $n \times m$ matrix, $\mathbb{W}^*$ will be $m \times n$. The product $\mathbb{W}^*\mathbb{W}$ is a symmetric matrix.

The quadratic measure of the smoothness can be defined as $\sum\limits_{j} (f_{j-1}-f_j)^2$ and is therefore given by Eq. (5.94), where $\mathbb{H}$ is near-diagonal and symmetric about the diagonal. Now we have to define a constraint, which selects F. A commonly used constraint is given by the variance

$$q = \sum_{j=1}^{n} (f_j-\bar{f})^2 \; , \quad \bar{f} \text{ being the average} \quad n^{-1} \sum_{j=1}^{n} f_j \; .$$

For this choice of a constraint the matrix $\mathbb{H}$ for Eq. (5.94) takes the following shape (Twomey 1977 b):

$$\mathbb{H} = \frac{1}{n} \begin{pmatrix} (n-1) & -1 & \dots & -1 \\ -1 & (n-1) & \dots & -1 \\ \vdots & & & \\ -1 & -1 & \dots & (n-1) \end{pmatrix} \equiv n^{-1} \| n\delta_{ij} - 1 \| , \qquad (5.95)$$

where $\delta_{ij} = 1$ for $i = j$ and $\delta_{ij} = 0$ for $i \neq j$.

For a large $n \times n$ matrix, i.e. for a large number n of measurements $g(v_i)$, and an equally large number n function values $f(z_j)$ at quadrature points z_j, the main diagonal contains values $1 - 1/n$ very close to one, and the off-diagonal values almost vanish as $1/n$.

Now we can proceed towards the solution of the inversion problem. We have to minimize Eq. (5.94), subject to the constraint (5.93), for which the inequality can be replaced by an equality, and the limit can be taken according to Eq. (5.93). This constrained extremum problem can be solved by finding an absolute (unconstrained) extremum of

$$(\mathbb{W}F - G)^* (\mathbb{W}F - G) + \Gamma F^* \mathbb{H} F ,$$

where Γ is still an undetermined multiplier. We therefore require that for all components k

$$\frac{\partial}{\partial f_k} [F^* \mathbb{W}^* \mathbb{W} F - G^* \mathbb{W} F - F^* \mathbb{W}^* G + \Gamma F^* \mathbb{H} F] = 0 .$$

When introducing the unit vectors $\partial F / \partial f_k = u_k$ and $\partial F^* / \partial f_k = u_k^*$ for the original and the transposed function vectors respectively, we get

$$u_k^* [\mathbb{W}^* \mathbb{W} F - \mathbb{W}^* G + \Gamma \mathbb{H} F] + [F^* \mathbb{W}^* \mathbb{W} - G^* \mathbb{W} + \Gamma F^* \mathbb{H}] u_k = 0 .$$

Bearing in mind that $\mathbb{H}^* = \mathbb{H}$, we find that the second bracketed term is the transpose of the first. This means that, if the sum vanishes, each term separately must vanish for components k. Hence $(\mathbb{W}^* \mathbb{W} + \Gamma \mathbb{H}) F - \mathbb{W}^* G = 0$ or

$$F = (\mathbb{W}^* \mathbb{W} + \Gamma \mathbb{H})^{-1} \mathbb{W}^* G , \qquad (5.96)$$

the equation for constrained linear inversion. The usual procedure for applying Eq. (5.96) is to choose several values of Γ, and to compute the residual $|\mathbb{W}^* F - G|$. If this is larger than the overall error in G due to quadrature and experimental errors, then Γ is too large and the solution has been constrained too much. If $|\mathbb{W}^* F - G|$ is smaller than the sum of all errors in G, one has an underconstrained solution. The choice of the most appropriate value for Γ, yielding the combined minimum of non-smoothness and of spurious oscillations, can be made on the basis of equality of the residual of $|\mathbb{W}^* F - G|$ and the sum of errors of G.

From the above discussion of the realization and accuracy of the inversion procedure, one might conclude that the dimension of the matrix $\mathbb{W}$ should be as high as possible, i.e. the interval (z_g, z_s) should be subdivided by as many quadrature points as feasible in terms of computing capacity. There is, however, another more fundamental limitation, which makes any further increase of n senseless beyond a certain value. This is dictated by the inter-dependence of the kernels (weighting functions) of the integral equation (5.81). When one of n measurements can be predicted mathematically from the others within an uncertainty given by the accuracy of the measurement, no significant increase of information is obtained by this measurement. The inter-dependence of the measurements allows one to represent the m-th measurement as a linear combination of all other $i \neq m$ measurements

$$g_m = \sum_{i \neq m} a_i g_i + \Delta_m \;,$$

with an uncertainty Δ_m. If this uncertainty of the mathematical representation is less than the inaccuracy of the measurements, we have redundancy. Mathematically expressed

$$|\Delta|^2 = \sum_i |\varepsilon_i|^2$$

is the upper bound for Δ to have redundancy, where ε_j are the relative measurement errors. If the measurement errors are randomly distributed, this bound can be written as

$$|\Delta|^2 \leqslant n|\varepsilon|^2 \;,$$

where ε is the r.m.s. error. Since $|\Delta|$ increases as $\sqrt{n}$, there is always improvement if n is increased, but this improvement will also be present even if previous measurements are simply repeated. This $\sqrt{n}$ improvement is caused by the increase in measurement accuracy obtained by repetition of measurements or by increasing the integration time of a measurement.

The same arguments of interdependence can be applied to kernels: if a kernel $w_l(z)$ can be expressed exactly as a weighted sum of the others, then a linear combination is possible, yielding

$$\sum_i a_i w_i = 0 \tag{5.97}$$

(with at least one $a_i \neq 0$) and one kernel – say w_l – can be expressed as

$$w_l(z) = -\frac{1}{a_l} \sum_{i \neq l} a_i w_i(z) \;.$$

The exact fulfillment of this linear combination is a sufficient and necessary condition for some g, say g_l, to be exactly predictable from the other g's

$$g_l = -\frac{1}{a_l} \sum_{i \neq l} a_i \int w_i(z)\, f(z)\, dz = -\frac{1}{a_l} \sum_{i \neq l} a_i g_i \;. \tag{5.98}$$

Equations (5.97) will rarely be exactly fulfilled, but there will always be a small error

$$\Delta_l(z) = w_l(z) - \sum_{i \neq l} \left(-\frac{a_i}{a_l}\right) w_i(z) = \frac{1}{a_l} \sum_i a_i w_i(z) \;. \tag{5.99}$$

Therefore, the condition (5.97) has to be modified to become a system of equations for the eigenvalue λ, as was formulated in (5.91), in order to find the optimum set of coefficients a_i. From this we can conclude

$$| \Delta_l(z) | = \frac{1}{|a_l|} \left| \sum_{i=1}^{n} a_i w_i(z) \right| = \frac{\sqrt{\lambda}}{|a_l|} . \tag{5.100}$$

Since a_l is at least $n^{-1/2}$, the square norm of the error in $w_l(z)$ is $\sqrt{n\lambda}$ or less.

Discussion of the error in g_i can be treated equally simply if the problem is restricted by appropriate scaling so that the individual g_i are of the order of unity. Then the ε_i can be identified with relative errors and − if e is the r.m.s. relative error − we get $|\varepsilon|^2 = ne^2$.

To relate the prediction error in g_l to the measurement error, we have to compare the approximately fulfilled equation (5.98) with the exact relationship

$$g_l + \varepsilon_l = \int w_l(z) f(z) \, dz \ ,$$

letting $f(z)$ range through all admissible functions.

Passing over the discussion of the various components which appear in the prediction formula of g_l (Twomey 1977b), we come to the concluding statement on the independence of the measurements g_i, which is closely related to (5.100): The independence of n measurements in the presence of a relative error of measurement $|\varepsilon|$ is assured, if

$$\lambda_{\min} > \frac{|\varepsilon|^2}{n} \ , \tag{5.101}$$

provided the system is properly scaled. Here $\lambda_{\min}$ is the lowest absolute eigenvalue. In general, if m eigenvalues are less than $|\varepsilon|^2$, then there are m redundant measurements which can be predicted from the $n-m$ measurements.

In the second half of this section we have touched on the problem of inverting remotely sensed data, starting from the situation encountered in atmospheric sounding by satellite-borne sensors. Other geometric situations pose even more problems of inversion. Ground-based sensors receiving the radiation as a function of elevation angle need a difference weighting function for retrieval (Randegger 1980). The most sensitive method, the limb-sounding configuration, i.e. looking from a satellite tangentially through the atmosphere, has the advantage of a long path of emitting molecules against a cold sky background. Infrared limb sounders (Gille et al. 1975) have been successfully applied on spacecraft for temperature and composition measurements (Gille et al. 1980), and satellite-borne microwave limb sounders have been proposed and are presently under construction (e.g. Schanda et al. 1976 and 1986). This method is distinguished by very narrow weighting functions.

Most of this text on inversion is a short summary of a part of Twomey's work (1977a, b). Much more literature about this complicated, but crucially important topic in the processing of remote sensing data exists, e.g. reviews such as by Deepak (1977), Houghton et al. (1984), and different approaches are put forward, such as the iterative method by Chahine (1972 and 1977), and Chahine and W. L. Smith (1983), as well as the a priori linear statistical method by Westwater and Strand (1968) and Rodgers (1976) or the multidimensional retrievals by Nathan et al. (1984).

Let us conclude this section with two examples which, although not results of an inversion procedure, however demonstrate the importance of a reliable inversion and of scrupulous application of radiative transfer.

Figure 5.14 presents the spectral radiance in the infrared range between about 6.6 μm (wave number 1500 cm^{-1}) and 25 μm (400 cm^{-1}) at three

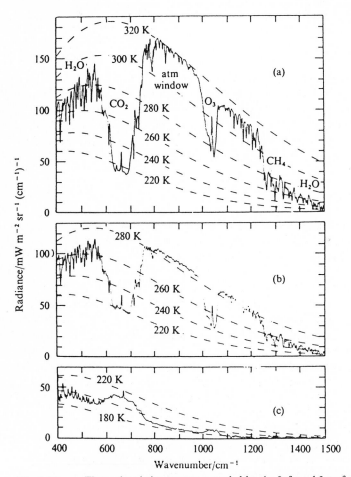

Fig. 5.14a – c. Thermal emission spectra recorded by the Infrared Interferometer Spectrometer IRIS, on Nimbus 4. Radiances of blackbody temperatures are superimposed. **a** Sahara; **b** Mediterranean; **c** Antarctic. (Hanel et al. 1971)

climatically very different locations on the earth, in particular concerning soil temperature. The temperature of the upper tropospheric air − when averaged over the width of the weighting function − does not vary much with latitude. Therefore the CO_2-absorption band shows about the same temperature at all latitudes. The atmospheric window regions show the soil temperatures.

Figure 5.15 represents the albedo of the globe as a function of latitude. The absorbed solar energy over latitude is exactly the inverted albedo in the visible part of the spectrum. The energy emitted by the earth depends more on temperature than on latitude, and of course on the albedo in the thermal infrared range. The earth absorbs energy in the zones of small albedo in the visible, depending on the incidence angle of the sun, thanks to the atmospheric transparency in this part of the spectrum. It loses energy rather uniformly over the globe due to the partial transparency in the thermal infrared. The extraordinary importance of water in the heat budget of the earth is obvious. For steep solar incidence, i.e. in the tropical region, the albedo of water is very low, hence absorption of the sun's radiation is high.

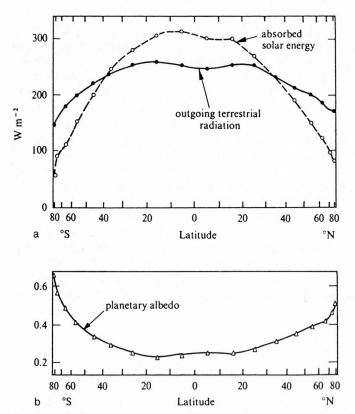

Fig. 5.15. a Solar energy absorbed and terrestrial radiative energy emitted by the earth atmosphere system. **b** Earth's albedo measured by satellite. (Von der Haar and Suomi 1971)

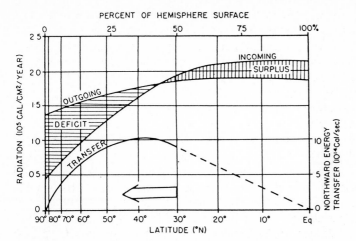

Fig. 5.16. Meridional transport of energy (Copyright 1964 by Scientific American Inc.)

The polar ice caps with high albedo reflect the energy influx incident in the visible spectrum, however, their lower albedo in the infrared causes effective emission of thermal radiation. The strong dependence of the emitted radiation on temperature (Stefan-Boltzmann's T^4-law) balance the earth's temperature to a well-defined mean and limits the ocean surface temperature to about 30 °C. Not only the surface with its various albedos affects the global heat budget, but additionally the atmosphere presents an important and highly variable intervening medium. Water vapour causes wavelength-selective absorption and re-emission of radiation, and clouds cause efficient reflection of visible and infrared radiation due to scattering by droplets. Strong water vapour and cloud concentrations take place predominantly in the intertropical convergence zone. The unequal distribution in latitude of water in the solid, liquid and gaseous state causes an imbalance of the radiative transport of heat on earth (Fig. 5.15). This imbalance must be compensated by a poleward transfer of energy in the atmosphere. Figure 5.16 presents the yearly average of this imbalance and the consequent transfer (Iribarne and Cho 1980).

The question of energy budget in the atmosphere leads well beyond the scope of this text and it is treated in various textbooks (e.g. Houghton 1977, Bolle 1982). However, for the solution of these problems, remote sensing must play an important role.

References

Abramowitz M, Stegun IA (1970) Handbook of mathematical functions, 9th edn. Dover Publ, New York

Ackerman M (1976) Measurements of minor constituents in the stratosphere. In: Burger JJ, Pedersen A, Battrick B (eds) Atmospheric physics from spacelab. Reidel, Dordrecht Boston

Allen WA, Richardson AJ (1968) Interaction of light with a plant canopy. J Opt Soc Am 58:1023 – 1028

Ballhausen CJ, Gray HB (1964) Molecular orbital theory. Benjamin, New York Amsterdam

Baltes HP (ed) (1978) Inverse source problems in optics. Topics in current physics. Springer, Berlin Heidelberg New York

Baltes HP (ed) (1980) Inverse scattering problems in optics. Topics in current physics. Springer, Berlin Heidelberg New York

Barrick DE (1970) Rough surfaces. In: Ruck GT (ed) Radar cross section handbook, vol II. Plenum Press, New York London, p 671

Becker GE, Autler SH (1946) Water vapor absorption of electromagnetic radiation in the centimeter wavelength range. Phys Rev 70:300 – 307

Beckmann P, Spizzichino A (1963) The scattering of electromagnetic waves from rough surfaces. Pergamon Press, Oxford

Bingel WA (1967) Theorie der Molekülspektren. Verlag Chemie, Weinheim

Boerner WM, Brand H, Craw LA, Gjessing DT, Jordan AK, Keydel W, Schwierz G, Vogel M (eds) (1985) Inverse methods in electromagnetic imaging, parts I and II. Reidel, Dordrecht Boston

Boettcher CJF, Bordewijk P (1973 vol 1, 1978 vol 2) Theory of electric polarization. Elsevier North-Holland, Amsterdam

Bolle HJ (1982) Radiation and energy transport in the earth atmosphere system. In: Hutzinger O (ed) Handbook of environmental chemistry, vol 1 part B. Springer, Berlin Heidelberg, p 131

Born M (1965) Optik, 1st edn 1933. Springer, Berlin Heidelberg New York

Born M, Wolf E (1964) Principles of optics, 2nd edn. Pergamon Press, Oxford

Brasseur G, Solomon S (1984) Aeronomy of the middle atmosphere. Reidel, Dordrecht Boston

Bristow MPF, Bundy DH, Edmonds CM, Ponto PE, Frey BE, Small LF (1985) Airborne laser fluorosensor survey of the Columbia and Snake rivers simultaneous of chlorophyll, dissolved organics and optical attenuation. Int J Remote Sensing 6:1707 – 1734

Brown GS (1984) The validity of shadowing corrections in rough surface scattering. Rad Sci 19:1461 – 1468

Chahine MT (1972) A general relaxation method for inverse solution of the full radiative transfer equation. J Atmos Sci 29:741

Chahine MT (1977) Generalization of the relaxation method for the inverse solution of nonlinear and linear transfer equations. In: Deepak A (ed) Inversion methods in atmospheric remote sounding. Academic Press, London New York, p 67

Chahine MT, Smith WL (1983) J Atmos Sci 40:2025 – 2035

Chan K, Ito H, Inaba H, Furuya T (1985) 10 km-long fibre-optic remote sensing of CH_4 gas by near infrared absorption. Appl Phys B 38:11 – 15

Chandrasekhar S (1960) Radiative transfer. Dover Publ, New York

Chu TS, Hogg DC (1968) Effects of precipitation on propagation at 0.63, 3.5 and 10.6 microns. Bell Syst Techn J 47:723 – 759

Cole RH (1961) Theories of dielectric polarization and relaxation. In: Birks JB, Hart J (ed) Progress in dielectrics, vol III. Heywood, London

Collis RTH, Russel PB (1976a) Laser applications in remote sensing. In: Schanda E (ed) Remote sensing for environmental sciences. Springer, Berlin Heidelberg New York

Collis RTH, Russel PB (1976b) Lidar measurement of particles and gases by elastic back-scattering and differential absorption. In: Hinkley ED (ed) Laser monitoring of the atmosphere. Springer, Berlin Heidelberg New York

Coulson KL (1975) Solar and terrestrial radiation. Academic Press, London New York

Croom DL (1978) Millimetre remote sensing of the stratosphere and mesosphere. In: Lund T (ed) Surveillance of environmental pollution and resources by electromagnetic waves. Reidel, Dordrecht Boston

Debye P (1929) Polar molecules, chap V. Chem Catalog Co, New York

Deepak A (ed) (1977) Inversion methods in atmospheric remote sounding. Academic Press, London New York

Deirmendjian D (1969) Electromagnetic scattering on spherical polydispersions. Am Elsevier, New York

Fabian J, Hartmann H (1980) Light Absorption of organic colorants. In series: Reactivity and structure concepts in organic chemistry, vol XII. Springer, Berlin Heidelberg New York

Farrow JB (1975) The influence of the atmosphere on remote sensing measurements. Eur Space Agency Rep CR 354, Neuilly

Fletcher NH (1962) The physics of rain clouds. Cambridge Univ Press, Cambridge

Fletcher NH (1970) The chemical physics of ice. In: Monographs on physics. Cambridge Univ Press, Cambridge

Fraser KS, Gault NE, Reifenstein EC, Sievering H (1975) Interaction mechanisms within the atmosphere. In: Reeves RG (ed) Manual of remote sensing, vol I. Am Soc of Photogrammetry, Falls Church, Virginia, p 181

Fung AK, Chen MF (1984) The effect of wavelength filtering in rough surface scattering. Proc Int Geosci Rem Sens Symp, Strasbourg, p 631

Gausman HW, Allen WA, Wiegand CL, Escobar DE, Rodriguez RR (1971) Leaf light reflectance, transmittance, absorption, and optical and geometrical parameters for eleven plant genera with different leaf mesophyll arrangements. In: Proc 7th Symp Remote Sens Environ, Univ Michigan, Ann Arbor, p 1599

Gille JC, Bailey PL, House FB, Craig RA, Thomas JR (1975) In: Sissala JE (ed) Nimbus-6 users guide. Nat Aeronaut Space Administ, Greenbelt Maryland

Gille JC, Bailey PL, Russel JM (1980) Temperature and composition measurements from LRIR and LIMS experiments on Nimbus 6 and 7. Philos Trans R Soc London Ser A 296:205 – 218

Goetz AFH (1985) Imaging spectrometry from aircraft and space platforms, Internat. Coll. Spectral Signatures of Objects in Remote Sensing, Les Arcs, 95

Gordy W, Cook RL (1970) Microwave molecular spectra. Wiley Interscience, New York London

Griffiths J (1976) Colour and constitution of organic molecules. Academic Press, London New York

Groebner W, Hofreiter N (1966) Integraltafel II, Bestimmte Integrale. Springer, Wien New York

Haar TH von der, Suomi VE (1971) Measurements of the earth's radiation budget from satellites during a five year period. Part I: Extended time and space means. J Atmos Sci 28:305

Hake RD, Arnold DE, Jackson DW, Evans WE, Ficklin BP, Long RA (1972) Dye-Laser observations of the night-time atomic sodium layer. J Geophys Res 77:6839 – 6848

Haken H (1981) Light, vol 1: Waves, photons, atoms. Elsevier North Holland, Amsterdam New York Oxford

Hanel RA, Schlachman B, Rogers D, Vanous D (1971) Nimbus-4 Michelson interferometer. Appl Opt 10:1376 – 1382

Hasted JB (1961) The dielectric properties of water. In: Birks J, Hart J (ed) Progress in dielectrics, vol III. Wiley, New York

Hasted JB (1974) The dielectric properties of water. In: Luck WAP (ed) Structure of water and aqueous solutions. Verlag Chemie, Weinheim, p 377

Hellwege AM (1962) Brechzahl reiner anorganischer Fluessigkeiten. In: Hellwege KH, Hellwege AM (eds) Landolt-Boernstein, part 8. Springer, Berlin, p 5 – 569

Herzberg G (1945) Molecular spectra and molecular structure. Reinhold, New York London

Hippel AR von (1954) Dielectrics and waves. M.I.T. Press, Mass Inst Technol, Cambridge Mass
Hoekstra P, Spanogle D (1972) Radar cross section measurements of snow and ice. Cold Regions
    Research and Engineering Lab CRREL – Techn Rep 235, Hanover New Hampshire, USA,
    p 4
Hoffer RM, Johannsen CJ (1969) Ecological potential in spectral signature analysis. In: Remote
    sensing in ecology. Univ Georgia Press, Athens Georgia, p 1
Houghton JT (1977) The physics of atmospheres. Cambridge Univ Press, Cambridge
Houghton JT, Taylor FW, Rodgers CD (1984) Remote sounding of atmospheres. Cambridge
    Univ Press, Cambridge
Howarth O (1973) Theory of spectroscopy. Wiley, New York
Hudson RD (1969) Infrared system engineering. Wiley, New York
Hulst HC van de (1957) Light scattering by small particles. Wiley, New York
Inaba H, Kobayashi T (1972) Laser-raman radar. Opto-Electronics 4:101 – 123
Iribarne JV, Cho HR (1980) Atmospheric physics. Reidel, Dordrecht Boston
Ishimaru A (1978) Wave propagation and scattering in random media. Academic Press, London
    New York
Joss J, Gori EG (1978) Shapes of raindrop size distributions. J Appl Meteorol 17:1054 – 1061
Kamen MD (1963) Primary processes in photosynthesis. Academic Press, London New York
Kerker M (1969) The scattering of light and other electromagnetic radiation. Academic Press,
    London New York
Kittel C (1967) Introduction to solid state physics, 3rd edn. Wiley, New York
Laws JO, Parsons DA (1943) The relation of raindrop size to intensity. Am Geophys Un Trans
    24:452 – 460
Leader JC (1979) Analysis and prediction of laser scattering from rough-surface materials. J Opt
    Soc Am 69:610 – 628
Liebe HJ (1981) Modeling attenuation and phase of radiowaves in air at frequencies below
    1000 GHz. Radio Sci 16:1183 – 1199
Litvin FF, Sineshchekov VA (1975) Molecular organization of chlorophyll and energetics of the
    initial stages of photosynthesis. In: Govindjee (ed) Bioenergetics of photosynthesis. Academic
    Press, London New York, p 619
MacDonald DKC (1962) Noise and fluctuations: an introduction. Wiley, New York
Maetzler C, Ramseier R, Svendsen E, Farrelly B, Horjen I (1980) Norsex Report II, 2nd edn.
    Geophys Inst Univ Bergen Norway
Marshall JS, Palmer WMK (1948) The distribution of raindrops with size. J Meteorol 5:165 – 166
McClatchey RA, Selby JEA (1972) Atmospheric transmittance 7 – 30 μm. Airforce Cambridge
    Res Lab Environ Res Pap No 419. LG Hanscom Field, Mass
McClatchey RA, Benedict WS, Clough SA, Burch DE, Calfee RF, Fox K, Rothman LS, Garing
    JS (1973) AFCRL Atmospheric absorption line parameters compilation. In: Airforce
    Cambridge Res Lab Environ Res Pap No 434. LG Hanscom Field, Mass
Meier H (1963) Die Photochemie der organischen Farbstoffe. Springer, Berlin Goettingen
    Heidelberg
Moore RK (1983) Radar fundamentals and scatterometers. In: Colwell RN, Simonett DS, Ulaby
    FT (eds) Manual of remote sensing, vol I, 2nd edn. Am Soc Photogrammetry, Falls Church,
    Virgina, pp 369 – 427
Myers VI (1975) Crops and soils. In: Reeves RG, Anson A, Landen D (eds) Manual of remote
    sensing, vol II. Am Soc Photogrammetry, Falls Church, Virginia, p 1715
Nathan KS, Rosenkranz PW, Staelin DH (1984) Abstr Nat Radio Sci Meet, 11 – 13 Jan 1984,
    Boulder Colorado, p 164
NORSEX Group (1983) Norwegian remote sensing experiment in a marginal ice zone. Science
    220:781 – 787
Norwood VT, Lansing JC (1983) Electro-optical imaging sensors. In: Colwell RN, Simonett DS ,
    Ulaby FT (eds) Manual of remote sensing, vol I, 2nd edn. Am Soc Photogrammetry, Falls
    Church, Virginia, pp 335 – 367
Peacock TE (1972) The electronic structure of organic molecules. Pergamon Press, Oxford

Penner SS (1959) Quantitative molecular spectroscopy and gas emissivities. Addison-Wesley, Reading, Mass, Pergamon Press, Oxford

Poynter RL, Pickett HM (1984) Submillimeter, millimeter and microwave spectral line catalogue. Jet Propulsion Lab Publ 80 – 23, Rev 2, Pasadena, Calif

Pruppacher HR, Klett JD (eds) (1978) Microphysics of clouds and precipitation. Reidel, Dordrecht Boston

Randegger AK (1980) On the determination of the atmospheric ozone profile for ground based microwave measurements. Pure Appl Geophys 118:1052 – 1065

Ray PS (1972) Broadband complex refractive indices of ice and water. Appl Opt 11:1836 – 1844

Robinson BF, de Witt DP (1983) Electro-optical non-imaging sensors. In: Colwell RN, Simonett DS, Ulaby FT (eds) Manual of remote sensing, 2nd edn. Am Soc Photogrammetry, Falls Church, Virginia, p 293

Rodgers CD (1976) Rev Geophys Space Phys 14:609 – 624

Schanda E (1971) Graphs for radiometer applications. Elektron Telecommun 4:51 – 55

Schanda E (1976) Passive microwave sensing. In: Schanda E (ed) Remote sensing for environmental sciences. Springer, Berlin Heidelberg, p 187

Schanda E, Fulde J, Kuenzi K (1976) Microwave limb sounding of strato- and mesosphere. In: Burger JJ, Pedersen A, Battrick B (eds) Atmospheric physics from spacelab. Reidel, Dordrecht Boston, p 135 – 146

Schanda E, Künzi K, Kämpfer N, Hartmann G, Degenhart W, Keppler E, Loidl A, Umlauft G, Vasyliunas V, Zwick R, Schwartz PR, Bevilaqua RM (1986) Millimeter wave atmospheric sounding from Space Shuttle. Acta Astronautica 13 (in press)

Sellers PJ (1985) Canopy reflectance, photo-synthesis and transpiration. Internat. J Remote Sensing 6:1335 – 1372

Shaw JH, Chahine MT, Farmer CB, Kaplan LD, McClatchey RA, Schaper PW (1970) Atmospheric and surface properties from spectral radiance observations in the 4.3 micron region. J Atmos Sci 27:301 – 306

Siegert AJ, Ridenour LN, Johnson MH (1963) Properties of radar targets. In: Ridenour LN (ed) Radarsystem engineering. Boston Techn Lithogr, Lexington, Mass

Slater PN, Doyle FJ, Fritz NL, Welch R (1983) Photographic systems for remote sensing. In: Colwell RN, Simonett DS, Ulaby FT (eds) Manual of remote sensing, vol I, 2nd edn. Am Soc Photogrammetry, Falls Church, Virginia, p 231

Smith RA, Jones FE, Chasmar RP (1968) The detection and measurement of infra-red radiation, 2nd edn. Clarendon Press, Oxford

Smith WL (1970) Iterative solution of the radiation transfer equation for the temperature and absorbing gas profile of an atmosphere. Appl Opt 9:1993

Sokolnikoff IS, Redheffer RM (1958) Mathematics of physics and modern engineering. McGraw Hill, New York London

Sommerfeld A (1948) Elektrodynamik. Dietrich, Wiesbaden

Staelin DH (1977) Inversion of passive microwave remote sensing data from satellites. In: Deepak A (ed) Inversion methods in atmospheric remote sounding. Academic Press, London New York, p 361

Staelin DH (1983) Atmospheric temperature sounding from space. Abstr 9th Conf Aerospace Aeronaut Meteorol. Am Meteorol Soc, June 6 – 9, Omaha, Nebr, pp 191 – 193

Staelin DH, Kunzi KF, Pettyjohn RL, Poon RKL, Wilcox RW (1976) Remote sensing of atmospheric water vapor and liquid water with the Nimbus 5 microwave spectrometer. J Appl Meteorol 15:1204 – 1214

Stratton JA (1941) Electromagnetic theory. McGraw Hill, New York London

Svanberg S (1978) Fundamentals of atmospheric spectroscopy. In: Lund T (ed) Surveillance of environmental pollution and resources by electromagnetic waves. Reidel, Dordrecht Boston, p 37

Tiuri ME (1966) Radio telescope receivers. In: Kraus JD (ed) Radio astronomy. McGraw Hill, New York, p 236

Townes CH, Schawlow AL (1955) Microwave spectroscopy. McGraw Hill, New York London

Tsang L, Kong JA, Shin RT (1985) Theory of microwave remote sensing. Wiley, New York

Twomey S (1977a) Some aspects of the inversion problem in remote sensing. In: Deepak A (ed) Inversion methods in atmospheric remote sounding. Academic Press, London New York, p 41

Twomey S (1977b) Introduction to the mathematics of inversion in remote sensing and indirect measurements. Elsevier Sci Publ, Amsterdam Oxford New York

Ulaby FT, Batlivala PP, Dobson MC (1978) Microwave backscatter dependence on surface roughness, soil moisture and soil texture, part II: Vegetation covered soil. IEEE Trans Geosci Electron 17:33–40

Ulaby FT, Moore RK, Fung AK (1981) Microwave remote sensing, vol I. Addison Wesley, Reading, Mass

Ulaby FT, Moore RK, Fung AK (1982) Microwave remote sensing, vol II. Addison Wesley, Reading, Mass

Valley SL (ed) (1965) Handbook of geophysics and space environment. Airforce Cambridge Res Lab, Cambridge, Mass

Waters JW (1976) Absorption and emission of microwave radiation by atmospheric gases. In: Meeks ML (ed) Methods of experimental physics, vol XII, part B. Radio astronomy. Academic Press, London New York, p 142

Waters JW, Staelin DH, Kunzi KF, Pettyjohn RL, Poon RKL (1975a) Microwave remote sensing of atmospheric temperatures from the Nimbus 5 satellite. COSPAR Space Research XV. Akademie Verlag, Berlin, p 117

Waters JW, Pettyjohn RL, Poon RKL, Kunzi KF, Staelin DH (1975b) Remote sensing of atmospheric temperature profiles with the Nimbus 5 microwave spectrometer. J Atmos Sci 32:1953–1969

Waters JW, Hardy JC, Jarnot RF, Pickett HM, Zimmermann P (1984) A balloon-borne microwave limb sounder for stratospheric measurements. J Quant Spectrosc Radiat Transfer 32:407–433

Weast RC (ed) (1974) Handbook of chemistry and physics, 55th edn. Chem Rubber Comp Press, Cleveland, Ohio

Westwater ER, Strand ON (1968) Statistical information content of radiation measurements used in indirect sensing. J Atmos Sci 25:750–758

Wheatley PJ (1968) The determination of molecular structure. Oxford Univ Press, Oxford

Yeh C (1964) Perturbation approach to the diffraction of electromagnetic waves by arbitrarily shaped dielectric obstacles. Phys Rev 135:A1193–A1201

Subject Index

Wave function 64f.
 – number 12, 31
 – vector 13
 – velocity 12, 26
Wavefront reconstruction 4
Wavelenght 5
Waves 9f.
Weighting function 159f.

Width of spectral line 47
Wien's approximation 142
Wien's displacement law 37, 142

Zeeman effect 58
Zero-point energy 42
 – field 47
 – oscillations 47